**TABLE OF CONTENTS:**                          **Page No.**

# METAPHORICAL VIEW OF PLANETARY MOTIONS INSIDE THE SOLAR SYSTEM

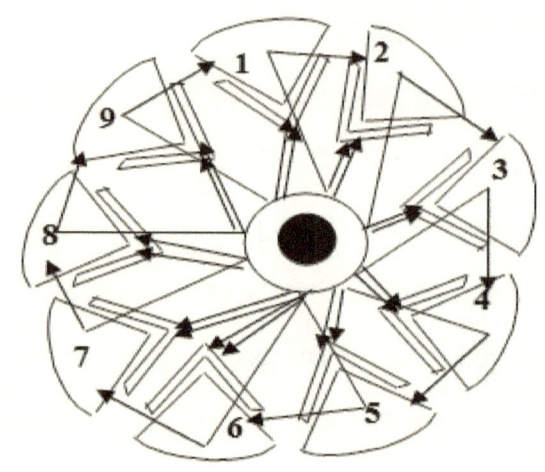

FIG.1

KEY

| | |
|---|---|
| ● | The Sun |
| 1 | Mercury |
| 2 | Venus |
| 3 | Earth |
| 4 | Mars |
| 5 | Jupiter |
| 6 | Saturn |
| 7 | Uranus |
| 8 | Neptune |
| 9 | Pluto |

A source of light [Bulb]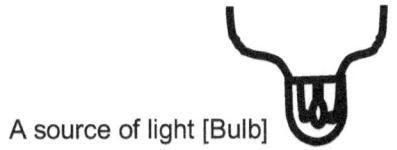

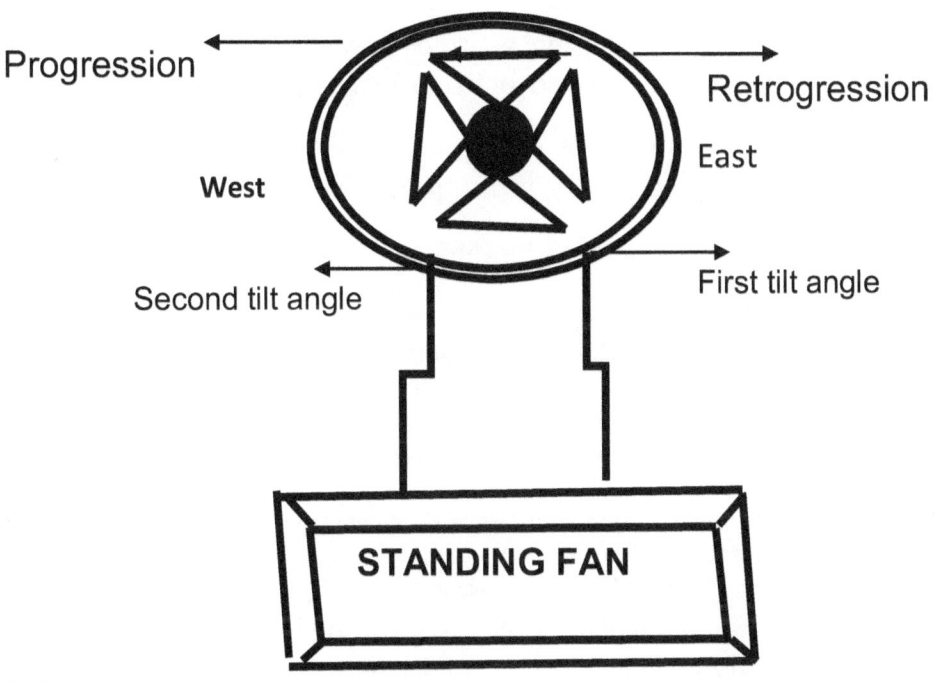

Progression

Retrogression

East

West

Second tilt angle

First tilt angle

STANDING FAN

FIG.2

## FIG.1: <u>The First To Enter Night Would Be The First To Enter Morning:</u>

-The outward motion [retrogression] would be pushing the East away from the SUN, but bringing the West closer to the SUN.

-The inward motion [progression] would be pushing the West away from the SUN, but bringing the East closer to the SUN.

-       Therefore, Retrogression is fast – forwarding, while Progression is rewinding

## FIG.2: <u>The First To Enter Night Would Be The First To Enter Morning:</u>

-The outward motion [retrogression] would be pushing the East away from the light bulb, but bringing the West closer to the light bulb.

-The inward motion [progression] would be pushing the West away from the light bulb, but bringing the East closer to the light bulb.

-Therefore, Retrogression is fast – forwarding, while Progression is rewinding.

### *SCIENTIFIC CONTINUUM*
The 'Contraction and Expansion' of the world of science is highly influenced by two intellectual forces, namely   "Discordance" and "Concordance"
The uniqueness of science is equally attributed to these forces of nature through all the ocarinas of time: decade by decade, century by century, millennium by millennium.
Hence, Expansion has no superiority over Contraction, because the later is the precursor of the former, but not the opposite.
These two intellectual forces can be viewed through the premises of physics to mean "Repulsion and Attraction", "Retrogression and Progression" respectively. Therefore, the first wonder of science is here summarized in these four related hypotheses: How does "Repulsion Give Rise to Attraction"? How does "Retrogression Give Rise to Progression"? How does "Contraction Give Birth to Expansion"? How does "Discordance Give Birth to Concordance?

## BACKGROUND OF THE STUDY

The motion of planets in the solar system is a phenomenon which remained something of a curiosity among scientists since antiquity. For example: Why do the planets closer to the sun tend to revolve faster round the sun, but rotate slowly on their various axes? Also, why do the planets that are far away from the sun tend to rotate faster on their various axes, but revolve slowly round the sun? Why does weight varies across the planets? Why do celestial bodies work with different clocks? Where is the origin of gravitational force? Where can we find gravitational field? You may think of Newton's view of gravitation, but you would not find complete answer to the questions at hand. You may also think of Einstein's view of gravitation, but the answer would not be different from the one above. Hence, there is need to research further to find posible answers to some of the above questions. This guide provides answers to some of these questions using an approach that is direct and standard

## A BRIEF REVIEW OF "BIG BANG" THEORY

The creation of the universe is explained by astrophysicists in a widely accepted phenomenon, popularly known as the *Big Bang*. It is supported by observational and experimental data gathered by astronomers and astrophysicists for decades.

According to the Big Bang, the whole universe was initially one Big Mass (Primary Nebula). Then there was a 'Big Bang' (Secondary separation) which resulted in the formation of galaxies, these then divided to form stars, planets, the sun, the moon etc. If all planetary bodies were evolved from the same source as theorized according to the Big Bang theory, then gravitational force should be a force of *Repulsion,* without which the entire universe would crumble away. The first forces of nature are always found together with weights. Hence, the foundation of gravitation was initially constructed upon '*Force and Weight*'.

## DESCRIPTION OF THE MACHINE

*Lingual-Numeric Calculator*

## DEFINTION

It is a three-dimensional formula that was designed to complement the work of modern computer in a way that is direct and precise for the advancement of human knowledge. It supports the Pythagorean assertion which says: "*All things had their origin and composition in numbers*".

It was allegorically expressed as:

## First Dimension

Input= [$O^I$ x Q] A Slide per Letter

Output In Figures=Cardinal Value
**Second Dimension**
Input=[1] + [$O^I$ x Q] A Slide per Letter
Output In Figures =First Ordinal Value
**Third Dimension**
Input=[$O^I$ x P] A Slide per Letter
Output In Figures=Second Ordinal Value
## FULL OF MEANING OF THE ABBREVIATIONS:
O-Stands for the <u>overall number</u> of English alphabets [26]
I-Stands for <u>indexes</u> (in descending order) to which the overall number of English alphabets will be multiply by itself. It depends on the number of letters inside a word
Q-Stands for the <u>quantitative value</u> of each letter of English when they are arranged in an alphabetic order
P-Stands for the <u>position</u> of each letter of English when they are arranged in an alphabetic order
For example:
## THE QUANTITATIVE VALUE OF EACH LETTER OF ENGLISH
*A=0,B=1,C=2,D=3,E=4,F=5,G=6,H=7,I=8,J=9,K=10,L=11,M=12*
*N=13,O=14,P=15,Q=16,R=17,S=18,T=19,U=20,V=21,W=22,X=23,Y=24*
*Z=25*
## THE POSITION OF EACH LETTER OF ENGLISH
*A=1,B=2,C=3,D=4,E=5,F=6,G=7,H=8,I=9,J=10,K=11,L=12,M=13N=14,O=15,*
*P=16,Q=17,R=18,S=19,T=20,U=21,V=22,W=23,X=24,Y=25,Z=26*


## ALGORITHM
### -INPUT
Input can be any word of interest. It can be a name of disease, term, course, discipline etc
### -DATA
Data is the representation of `a word in accordance with the three components of the   machine, namely: overall number, indexes and quantity
### -COMPUTING
Computing is the act of reading a word according to the computational system of the machine.
### -OUTPUT
Output is two: in figures and in words:
-Output in figures is the cardinal or ordinal value of an input.

-Output in words are vocabularies or sentences recognized by the machine during the process of interpretation of an input

## LANGUAGE OF THE MACHINE:
The basic difference betweencomputer and lingual-numeric calculator is the language of the machine: the former understands 0 & 1[mechanically: 'on' & 'off'], while the later understands1 & 1 [literally: 'relevant' or 'irrelevant' based on 'characters']

## MECHANISM OF DATA PROCESSING FOR AN OUPUT
Firstly, upon entering an input, the machine will ask you to select from one of the three dimensions of the formula. Secondly, it will derive an ordinal scale from the obtained value for segmented coding. Thirdly, the machine is going to compute the value obtained into one of the language of the machine for integrated coding. Fourthly, the machine will start working on various sets of database [e.g. f1, f2, f3, f4, etc] for output processing. It categorizes words into two supporting components, such as: vocabulary [one or more] and sentence [one or more].

Vocabulary- only the relevant words would be recognized by the machine, while the irrelevant words would not be recognized for lack of associated characters literally or metaphorically.

- Sentence- the machine would only recognize a sentence whose integrated code doesn't contradict the integrated code of the subject in a sentence and the general contents of the first component

## SEGMENTED CODE:
Is the act of considering all the numbers inside the cardinal or ordinal value of an input when making interpretation, it always requires an ordinal scale.

## INTEGRATED CODE:
Is the act of considering a number reduced into the language of the machine from the cardinal or ordinal value of an input when making interpretation. It can be dependent or independent.

For example-it is dependent when it was used to interpret an ordinal scale of other input, but it is independent when it was used to interpret itself.

## METHOD OF DATABASE PROCESSING USED BY THE MACHINE:
F1=[ 1-letter word database]
F2=[ 2-letter word database]
F3=[ 3-letter  word database]
F4=[ 4-letter word database]
F5=[ 5-letter word database] up to F22=[ 22-letter word database]
Hence, any word with more than 22 letters would be read as coined word whose morphemes will be computed separately, before joining into one word.

## INTRODUCTION TO THE MASTER CARD OF SOLAR SYSTEM

There are three astronomical eras in the history of astronomy, such as: prediction, estimation and observation era. These three astronomical eras revealed a chain of three great contributors in the study of Solar System, namely: astrologists, mathematicians and engineers.

From 15 to 17 century, the world of astronomy has made a remarkable achievement as our great founders like Galileo and Brahe discovered that the earth and other planets revolved around the sun. Kepler shows that planets move in an elliptical orbit, but not a circle, and Newton determined that gravitational forces exist between all the celestial bodies.

Similarly, from 19 to 21st century, computer and software engineers have made a significant contribution in the world of science by inventing the computer and the various types of memory chips which can be used to retrieve storage data from a computer without fear of missing the least. This modern technology has assisted scientists towards analyzing and clarifying their various sources of knowledge and information from one single multipurpose tool.

For example:
- *In Computer, there is a 'Memory Device'*
- *In Airplane, there is a 'Black Box'*
- *In Human Body, there is a 'Human Genome'*
- *In Solar System, there is 'Gravitation*

## DEFINITION OF SOLAR SYSTEM

The Solar System is the gravitationally bound system of the Sun and the objects that orbit it, either directly or indirectly. Of the objects that orbit the Sun directly, the largest are the eight planets, with the remainder being smaller objects, such as the five dwarf planets and small Solar System bodies. Of the objects that orbit the Sun indirectly—the moons—two are larger than the smallest planet, Mercury.

## MEMORY CARD OF SOLAR SYSTEM

DEFINITION- is a stepwise depiction of solar system using the vocabularies and sentences upon which the true understanding of gravitation depends on. Hence, it has two basic components, namely: vocabularies and sentences. It was derived using the lingual-numeric calculator.

### Critical Reasoning N0.1

Upon entering the word ' gravitation', the machine processed the output from various set of database at vocabulary level, such as: orbits, planets, cycles, radius, distance, weight, acceleration, mass, sun, galaxy, solar system, tropical year , core, supernova, infrared rays, rotation, light, atmosphere, helium, hydrogen, force, gravity, position, attraction, repulsion, vector, eclipse, emissions, magnetic poles, revolution, fusions, temperature, waves, gravitational field, metals, ellipse, axis, visible rays, gases, fire, gamma rays, motions, friction, momentum, gravitational force ,deflection, reflections, planetary crust, angle. You are to use these parameters to formulate the true master card of solar system that will enhance students' understanding of the subject matter. Use any method that doesn't contradict the basic rules of science.

Working From The First Dimension, We Have:

Input= [OI x Q] A Slide per Letter

**INPUT = GRAVITATION DATA**

O -26

I - 11 letters: 10, 9,8,7,6,5,4,3,2,1,0

Q-g = 6, r = 17, a =0, v = 21, i = 8, t =19, a = 0, t = 19, i =8, 0 = 14, n =13

COMPUTING

g[2610 x 6] + r[269 x 17] + a[268 x 0] + v[267 x 21] + i[266  x 8] +t[265 x 19] + a[264 x 0] +t [263   x 19] +i[262 x 8] + 0[261 x 14] + n[260 x 13]

**OUTPUT IN FIGURES**

[939,475,501,888,625]-this is the Cardinal Value of the word 'gravitation' inside the F11

Segmented Code

To derive the segmented code for the input, each of the definite number in the above cardinal value would appear only once.

For example:

[939,475,501,888,625]=[9-3-4-7-5-1-8-6-2]- Segmented Code

*It is this code that will be used to formulate the master card of solar system. It will consist of nine segments.*

> **OUTPUT IN WORDS**

 Topic No.1:

INPUT=CYCLES

 COMPUTING

c[265 x 2] + y[264 x 24] +c[263 x 2] +l[262 x 11] + e[261 x 4] + s[260 x 18]

OUTPUT

[34772886] -This is the Cardinal Value of the word 'cycles' inside the F6

Hence, to know the topic that the above word would fall inside the master card of solar system, the above cardinal value should be reduced into one of the language of the machine: [34772886 ]=[8+5+5+9+9+7+2]

[45]

[9] )-[9-3-4-7-5-1-8-6-2]
It implies that the word 'cycles' would fall under the first topic; hence it should form the primary knowledge of the subject matter.
INPUT=ORBITS
 COMPUTING
o[265 x 14] + r[264 x 17] +b[263 x 1] +i[262 x 8] + t[261 x 19] + s[260 x 18]
OUTPUT
[ 174131352]- This is the Cardinal Value of the word 'orbits' inside the F6
Hence, to know the topic that the above word would fall inside the master card of solar system, the above cardinal value should be reduced into one of the language of the machine:
[ 174131352 ]
[27]
[9] )-[9-3-4-7-5-1-8-6-2]
It implies that the word 'orbits' would fall under the first topic; hence it should form the primary knowledge of the subject matter.
Topic No. 2:
 INPUT=PLANETS
COMPUTING
p[266 x 15] +l[265 x 11] + a[264 x 0] +n[263 x 13] +e[262 x 4] + t[261 x 19] + s[260 x 18]
OUTPUT
[ 4764663480]- This is the Cardinal Value of the word 'planets' inside the F6
Hence, to know the topic that the above word would fall inside the master card of solar system, the above cardinal value should be reduced into one of the language of the machine:
[ 4764663480]
[48]
[12]
[3] )-[9-3-4-7-5-1-8-6-2]
It implies that the word 'planets' would fall under the second topic; hence it should form the primary knowledge of the subject matter.
INPUT=FRICTION
 COMPUTING
f[267 x 5] + r[266 x 17] +i[265 x 8] + c[264 x 2] +t[263 x 19] +i[262 x 8] + 0[261 x 14] + n[260 x
13]
OUTPUT
[ 45506923761 ]

10

[48]

[12]

[3] )-[9-3-4-7-5-1-8-6-2]

It implies that the word 'friction' would fall under the second topic; hence it should form the primary knowledge of the subject matter.

INPUT=GRAVITATIONAL FIELD

 COMPUTING

g[2612 x 6]+r[2611 x 17] + a[2610 x 0]+ v[269 x 21]+i[268 x 8]+ t[267 x 19]+a[266 x 0]+t[265 x 19]+i[264 x 8]+o[263 x 14]+n[262 x 13]+a[261 x 0]+l[260 x 11] /+ / f[264 x 5]+i[263 x 8]+e[262x 4]+ l[261 x 11]+d[260 x 3]

OUTPUT

[ 6350854392767710511]+[ 2428481 ]

[ 6350854392779138992 ]

[39]

[12]

[3] )-[9-3-4-7-5-1-8-6-2]

It implies that the word 'gravitational field' would fall under the second topic; hence it should form the primary knowledge of the subject matter.

INPUT=VISIBLE RAYS

 COMPUTING

v[266 x 21]+i[265 x 8]+s[264 x18]+i[263 x 8]+b[262 x 1]+l[261 x 11]+e[260 x 4] /+ / r[263 x 17]+a[262x 0]+ y[261 x 24]+s[260 x 18]

OUTPUT

[ 6590649446]+[ 299434 ]

[ 6590948880 ]

57]

[12]

[3] )-[9-3-4-7-5-1-8-6-2]

It implies that the word 'visible rays' would fall under the second topic; hence it should form the primary knowledge of the subject matter.

INPUT=LIGHT

 COMPUTING

l[264 x 11] +i[263 x 8] +g[262 x 6] + h[261 x 7] + t[260 x 19]

OUTPUT

[ 5171601 ]

[21]

[3] )-[9-3-4-7-5-1-8-6-2]

It implies that the word 'light' would fall under the second topic; hence it should

form the primary knowledge of the subject matter.
INPUT=WAVES
COMPUTING
w[264 x 22] +a[263 x 0] +v[262 x 21] + e[261 x 4] + s[260 x 18]
OUTPUT
[ 10067790 ]
[30]
[3] )-[9-3-4-7-5-1-8-6-2]
It implies that the word 'waves' would fall under the second topic; hence it should form the primary knowledge of the subject matter.
INPUT=REFLECTIONS
 COMPUTING
r[2610 x 17]+ e[269 x 4]+f[268 x 5]+ l[267 x 11]+e[266 x 4]+c[265 x 2]+t[264 x 19]+i[263 x 8]+o[262 x 14]+n[261 x 13]+s[260 x 18]
OUTPUT
[ 2422692394316940 ]
[66]
[12]
[3] )-[9-3-4-7-5-1-8-6-2]
It implies that the word 'reflections' would fall under the second topic; hence it should form the primary knowledge of the subject matter.
INPUT=DISTANCE
 COMPUTING
d[267 x 3] + i[266 x 8] +s[265 x 18] + t[264 x 19] +a[263 x 0] +n[262 x 13] + c[261 x 2] + e[260 x 4]
OUTPUT
[ 26789312892 ]
[57]
[12]
[3] )-[9-3-4-7-5-1-8-6-2]
It implies that the word 'distance' would fall under the second topic; hence it should form the primary knowledge of the subject matter.
INPUT=VECTOR
COMPUTING
v[265 x 21] + e[264 x 4] +c[263 x 2] +t[262 x 19] + o[261 x 14] + r[260 x 17]
OUTPUT
[251385177 ]
[39]
[12]

[3] )-[9-3-4-7-5-1-8-6-2]
It implies that the word 'vector' would fall under the second topic; hence it should form the primary knowledge of the subject matter.

Topic No.3:
INPUT=POSITION
COMPUTING
p[267 x 15] + o[266 x 14] +s[265 x 18] + i[264 x 8] +t[263 x 19] +i[262 x 8] + o[261 x 14] + n[260 x 13]
OUTPUT
[ 125019833809 ]
[49]
[13]
[4] )-[9-3-4-7-5-1-8-6-2]
It implies that the word 'position' would fall under the third topic; hence it should form the primary knowledge of the subject matter.
 INPUT=ANGLE
 COMPUTING
a[264 x 0] +n[263 x 13] + g[262 x 6] + l[261 x 11] + e[260 x 4]
OUTPUT
[ 232834]
[22]
[4] )-[9-3-4-7-5-1-8-6-2]
It implies that the word 'angle' would fall under the third topic; hence it should form the primary knowledge of the subject matter.
INPUT=RADIUS
 COMPUTING
r[265 x 17] + a[264 x 0] +d[263 x 3] +i[262 x 8] + u[261 x 20] + s[260 x 18]
OUTPUT
[ 202042066 ]
[22]
[4] )-[9-3-4-7-5-1-8-6-2]
It implies that the word 'radius' would fall under the third topic; hence it should form the primary knowledge of the subject matter.
INPUT=REVOLUTION
COMPUTING
r[269 x 17] + e[268 x 4] + v[267 x 21] + o[266 x 14] +l[265 x 11] + u[264 x 20] +t[263 x 19]
+i[262 x 8] + 0[261 x 14] + n[260 x 13]
OUTPUT

[ 93310003809841 ]

[40]

[4] )-[9-3-4-7-5-1-8-6-2]

It implies that the word 'revolution' would fall under the third topic; hence it should form the primary knowledge of the subject matter.

INPUT=ACCELERATION

 COMPUTING

a[2611 x 0] + c[2610 x 2]+ c[269 x 2]+e[268 x 4]+ l[267 x 11]+e[266 x 4]+r[265 x 17]+a[264 x 0]+t[263 x 19]+i[262 x 8]+o[261 x 14]+n[260 x 13]

OUTPUT

[ 294118294821169 ]

[67]

[13]

[4] )-[9-3-4-7-5-1-8-6-2]

It implies that the word 'acceleration' would fall under the third topic; hence it should form the primary knowledge of the subject matter.

Topic No.4:

INPUT=ZONE

 COMPUTING

z[263 x 25] + o[262 x 14] + n[261 x 13] + e[260 x 4]

OUTPUT

[ 449206 ]

[25]

[Z] )-[9-3-4-7-5-1-8-6-2]

It implies that the word 'zone' would fall under the fourth topic; hence it should form the secondary knowledge of the subject matter.

INPUT=PLANETARY CRUST

 COMPUTING

p[268 x 15]+ l[267 x 11]+a[266 x 0]+n[265 x 13]+e[264 x 4]+t[263 x 19]+a[262 x 0]+r[261 x

17]+y[260 x 24] + c[264 x 2]+r[263 x 17]+u[262 x 20]+s[261 x 18]+t[260 x 19]

OUTPUT

[ 3220912500778     ] +[ 1226751 ]

[ 3220913727529 ]

[52]

[7]-[9-3-4-7-5-1-8-6-2]

It implies that the word 'planetary crust' would fall under the fourth topic; hence it should form the secondary knowledge of the subject matter.

INPUT=METALS

COMPUTING

m[265 x 12]+e[264 x 4]+t[263 x 19]+a[262 x 0]+l[261 x 11]+s[260 x 18]

OUTPUT

[ 144738664 ]

[43]

[Z] )-[9-3-4-7-5-1-8-6-2]

It implies that the word 'metals' would fall under the fourth topic; hence it should form the secondary knowledge of the subject matter.

INPUT=ATTRACTION

COMPUTING

a[269 x 0]+t[268 x 19]+ t[267 x 19]+r[266 x 17]+a[265 x 0]+c[264 x 2]+t[263 x 19]+i[262 x

8]+o[261 x 14]+n[260 x 13]

OUTPUT

[ 4125571442161 ]

[43]

[Z] )-[9-3-4-7-5-1-8-6-2]

It implies that the word 'attraction' would fall under the fourth topic; hence it should form the secondary knowledge of the subject matter.

Topic No.5:

INPUT=HYDROGEN

COMPUTING

h[267 x 7]+y[266 x 24]+d[265 x 3]+r[264 x 17]+o[263 x 14]+g[262 x 6]+e[261 x 4]+n[260 x 13]

OUTPUT

[ 63680312813        ]

[50]

[5] )-[9-3-4-7-5-1-8-6-2]

It implies that the word 'attraction' would fall under the fifth topic; hence it should form the secondary knowledge of the subject matter.

INPUT=HYDROGEN

COMPUTING

h[267 x 7]+y[266 x 24]+d[265 x 3]+r[264 x 17]+o[263 x 14]+g[262 x 6]+e[261 x 4]+n[260 x 13]

OUTPUT

[ 63680312813        ]

[50]

[5] )-[9-3-4-7-5-1-8-6-2]

It implies that the word 'hydrogen' would fall under the filth topic; hence it

should form the secondary knowledge of the subject matter.
INPUT=HELIUM
COMPUTING
h[265 x 7]+e[264 x 4]+l[263 x 11]+i[262 x 8]+u[261 x 20]+m[260 x 12]
OUTPUT
[85196812 ]
[41]
[5] )-[9-3-4-7-5-1-8-6-2]
It implies that the word 'hydrogen' would fall under the fifth topic; hence it
should form the secondary knowledge of the subject matter.
INPUT=FUSIONS
COMPUTING
f[266 x 5]+u[265 x 20]+s[264 x 18]+i[263 x 8]+o[262 x 14]+n[261 x 13]+s[260 x
18]
OUTPUT
[ 1790582396 ]
[50]
[5] )-[9-3-4-7-5-1-8-6-2]
It implies that the word 'fusions' would fall under the fifth topic; hence it should
form the secondary knowledge of the subject matter.
INPUT=TEMPERATURE
 COMPUTING
t[2610 x 19]+ e[269 x 4]+ m[268 x 12]+ p[267 x 15]+e[266 x 4]+r[265 x
17]+a[264 x 0]+t[263 x 19]+u[262 x 20]+r[261 x 17]+e[260 x 4]
OUTPUT
[ 27065206720520 06 ]
[50]
[5] )-[9-3-4-7-5-1-8-6-2]
It implies that the word 'temperature' would fall under the fifth topic; hence it
should form the secondary knowledge of the subject matter.

GENERALLY, THE FIRST COMPONENT OF THE MEMORY   CARD OF SOLAR SYSTEM WILL BE DEPICTED AS:

## VOCABULARY-LEVEL

| • Primary Knowledge | | |
|---|---|---|
| Topic No.1 | Topic No.2 | Topic No. 3 |
| i-cycles | i-planets | i-position |
| ii-orbits | ii-friction | ii-angle |
| | iii-gravitational field | iii-radius |
| | iv-visible rays/light | iv-revolution |
| | v-waves | v-acceleration |
| | vi-reflections | |
| | vii-distance | |
| | viii-vector | |
| | ix-magnetic poles | |
| • Secondary Knowledge | | |
| Topic No. 4 | Topic No.5 | Topic No. 6 |
| i-zone | i-hydrogen | i-infrared rays/gamma rays |
| ii-planetary crust | ii-helium | ii-force |
| iii-metals | iii-fusions | iii-weight |
| iv-attraction | iv-temperature | iv-motions |
| | | v-momentum |
| | | vi-tropical year |
| • Tertiary Knowledge | | |
| Topic No.  7 | Topic No.8 | Topic No. 9 |
| i-core | i-mass | i-atmosphere |
| ii-iron | ii-ellipse | ii-milky way |
| iii-fire | iii-supernova | iii-galaxy |
| iv-repulsion | iv-axis | iv-solar system |
| v-gravity | v-eclipse | v-rotation |
| | | vi-gravitational force |
| | | vii-sun |
| | | viii-gases |
| | | ix-emissions |
| | | x-deflection |

## TRANSMITTERS OF GRAVITATIONAL FORCE

*Reading from the vocabulary component, gravitational force has three basic transmitters:*

*1- Emission of Gases 2-Infrared Rays/Gamma Rays 3- Visible Rays/Light*

## DEFINITION OF GRAVITATIONAL
## -NEWTON
According to the Newton, gravitational force is defined as the force of attraction which tries to pull two objects towards one another. In his universal law of gravitation, he said: "That every object attracts every other object with a force that, for any two objects, is directly proportional to the mass of each object and inversely proportional to the square of the distance between the two objects. His view of gravitational force was expressed mathematically as

$$Fgravity = \frac{Gm1\ m2}{r^2}$$

## -EINSTEIN
According to the Einstein's, gravitational force is defined as a distortion of space – time. In his gravitational principle, he said: "Matter tells space – time how to curve, and curve – space tells matter how to move. His view of gravitational force was expressed mathematically as:

$G\mu V = 8\lambda G/C^4 T\mu V$

## GMF DERIVATION
To come up with *new gravitational formula*, we need to sort out the basic parameters upon which the Newton & Einstein gravitational formulas were built and update them into parameters of *gravitational master formula*.

### - Newton's Gravitational Formula

$$Fgravity = \frac{Gm1\ m2}{r^2}$$

Basic parameters:
Gravity, Mass and Radius

### – Einstein's Gravitational Formula/Equation

$G\mu V = 8\lambda G/C^4 T\mu V$

Basic parameter:
Gravity, Velocity, Light and Time

### – Gravitational Master Formula
For planetary bodies e.g Mercury, Venus, Earth etc

$$GP = [\frac{GS}{AT} \times \frac{WP}{DP}]^{1/2}$$

For Stars, Moons etc.

$$GP = [\frac{GS}{AT} \times \frac{WP}{DP}]^{1/4}$$

Basic parameters:
Gravity, Weight, Distance and Time

## GMF DEDUCTION
-Gravity / Planetary Bodies

The first (Newton) and the second (Einstein) had emphasized on the importance of gravity to planetary bodies in their respective formulas, so also the third (GMF) has seen gravity from the same angle.

**Mass / Planetary Bodies**

The first had emphasized on the importance of mass to planetary bodies in his gravitational formula, but the second didn't see mass from the same angle, hence the third agrees with the first but with a slight difference, as "Mass" has been replaced by "Weight"

**Radius /Planetary Bodies**

The first had emphasized on the importance of radius to planetary bodies in his gravitational formula, but the second didn't see radius from the same angle, hence the third agrees with the first but with a slight difference, as "Radius" has been replaced by "Distance"[ from core to the surface of a planetary body.]

**Time / Planetary Bodies**

The second had emphasized on the importance of time to planetary bodies in his gravitational equations so also the third agrees with second on the subject matter holistically

**REAKDOWN OF THE GMF**

| FOR EARTH AND OTHER PLANETS |
| --- |
| $GP = \left[\frac{GS}{AT} \times \frac{WP}{DP}\right]\frac{1}{2}$ <br><br> Where: <br> $WP = \frac{WS}{LO}$ <br><br> $DP = \frac{DS}{LO}$ <br> $LO = \left[\frac{TY \times LT}{NO}\right]\frac{1}{2}$ <br><br> $TY = \left[\frac{LO}{\frac{1}{2}LT} \times NO\right]$ <br><br> $AT = \left[\frac{WP}{DP} \times SO\right]\frac{1}{2}$ <br><br> $SO = \left[\frac{NP \times NC}{SL} \times \frac{WS}{DP}\right]\frac{1}{2}$ <br> $AT1 = [TM1 - TM2]$ <br> Where: <br> $-TM1 = \left[\frac{WP}{DP} \times LO\right]\frac{1}{4}$ |

-TM2 = = $[\frac{WP}{DP} \times SO]\frac{1}{4}$

AT2= [TS-TL]
Where:

TS = $[\frac{DP}{NS \times OM}]\frac{1}{2}$

Note:
-DP = $\frac{DS}{SO}$
-NS = $\frac{WP}{GP}$
-WP = $\frac{WS}{SO}$
TL = $[\frac{DP}{NS \times OM}]\frac{1}{2}$
Note:
-DP = $\frac{DS}{LO}$
-NS = $\frac{WP}{GP}$
-WP = $\frac{WS}{LO}$

## FOR MOON AND OTHER STARS
GP = $[\frac{GS}{AT} \times \frac{WP}{DP}]\frac{1}{4}$
Where:
WP = $[\frac{WS}{LO}]\frac{1}{2}$
DP = $[\frac{DS}{LO}]\frac{1}{2}$
LO = $[\frac{AP \times LT}{NO}]\frac{1}{2}$
TY = $[\frac{LO}{\frac{1}{2}LT} \times NO]$
AT = $[\frac{WP}{DP} \times SO]\frac{1}{2}$
SO = $[\frac{NP \times NC}{SL} \times \frac{WS}{DP}]\frac{1}{4}$
$AT1 = [TM1 + TM2]$
Where
-TM1 = $[\frac{WP}{DP} \times LO]\frac{1}{4}$
-TM2 = = $[\frac{WP}{DP} \times SO]\frac{1}{4}$
AT2= [TS-TL]
Where:
TS = $[\frac{DP}{NS \times OM}]\frac{1}{12}$
Note:

-DP $=[\frac{DS}{SO}]1/_2$

-NS $=\frac{WP}{GP}$

- WP $=[\frac{WS}{SO}]1/_2$

TL $= [\frac{DP}{NS \times OM}]\frac{1}{12}$

Note :

-DP $= [\frac{DS}{LO}]$ ½

-NS $= \frac{WP}{GP}$

-WP $= [\frac{WS}{LO}]$ ½

## MISCELLANEOUS FORMS OF GMFFOR ALL HEAVENLY BODIES INSIDE THE SOLAR SYSTEM

i.RTL $= [\frac{OAR}{AL} \times OL \times NYM]$ Provided That RTL[ In thousand days]$-$
$\frac{TL}{OL}$[ $in\ thousand\ days$] =0

ii.TCH $= \frac{AT}{AM}$

iii.TCD$=\frac{TCH}{AH}$

iv. OL = [LO + SO]

v. From the relations: AT, LA & SA

-$[\frac{OL}{AL} =\frac{TCH}{(TS+TL)}]$for overall axial length

-$[\frac{LO}{LA} = \frac{TCH}{(TS+TL)}]$ for long axis

-$[\frac{SO}{SA} = \frac{TCH}{TS+TL}]$ for short axis

vi. STP $= [\frac{OL}{AL} X \frac{TS+TL}{AH}]$

Or

$[\frac{LO}{LA} X \frac{TS+TL}{AH}]$

Or

$[\frac{SO}{SA} X \frac{TS+TL}{AH}$

vii.CTR = (TS + TL) of a Planet much Closer To the sun
$\qquad$ (TS + TL) of a  planet far away from the sun

Note (TS + TL) most always be in round figure

viii..PSR = AT1 of planet (a) + AT1 of Planet (b) = AT1 of planet (c) provided that the three planets share some common features revolutionarily.

Also:

AT1 of planet (a), (b), (c) = AT1 of planet (d) provided that the four planets share some common features revolutionarily etc.

ix.   $PWD = [\frac{OL1}{OL2} \times \frac{1}{2}]$

x.   $PWN = [\frac{WP}{GP} \times AL]$

Or

$[\frac{WS}{GS} \times AL]$

xi. $TAR = [TCH \div \frac{LO}{SA}]$

xii. $EA = OAR - [ANC \times 360]$
Where:
$ANC = OAR/360$

## GMF STEPWISE APPLICATION

1-Find out the number of days taken by a planet to make a complete revolution (TY) in round figure e.g when the earth takes 365 days to make a complete revolution, the moon requires only 28 days carrying out the same task.

2-Substitute the values of TL, NO and TY inside the third unit of the formula to calculate (LO)

3-The values of (WS) should be divided by the value of (LO) in order to deduct the value of (WP)

4-The value of (DS) should also be divided by the value of (LO) in order to deduct the value of (DP)

5-Having the values of WP and DP at hand, you then proceed to calculate the value of (SO) since it is required before the value of (AT) can be calculated.

6-To calculate the value of (SO), you need to substitute the value of DP, OL, NP, NC and WS into the sixth unit of the formula.

7-Substitute the value of (SO) into the fifth unit of the formula to obtain the value of (AT)

8-Substitute the value of AT, GS, WP and DP into the general unit of the formula to obtain the value of GP

9-Use the values of WP, DP, LO and SO to calculate the value of TM1 & TM2 respectively, then proceed to calculate the value of (AT1)

10-Find out the value of TS & TL by substituting the values of OM,
DP $(\frac{DS}{SO}/\frac{DS}{LO})$, WP $(\frac{WS}{SO}/\frac{WS}{LO})$ and NS $(\frac{WP}{GP}/\frac{WP}{GP})$ respectively, then proceed to calculate the value of (AT2)

## FULL MEANING OF GMF ABBREVIATIONS

GP: gravitational force acting on a planet deducted from the gravitational force evolving from the core of the sun (in Newton)

GS: gravitational force generated inside the core of the sun (in metric tons)

WP: weight of a planet deducted from the weight of the sun (in metric tons)

DP: distance from the core to the surface of a planet deducted from the distance from the core to the surface of the sun (in thousand kilometers)

AT: actual time taken by a planet to make a complete revolution (in minutes)

WP: weight of the sun in metric tons.

DS: distance from the core to the surface of the sun

LO: long – orbital curvatures in thousand miles inside which a planet swim round the sun.

LO: long – orbital curvatures in thousand miles inside which a planet swim round the sun.

SO: short – orbital curvatures in thousand miles, inside which a planet swim round the sun.

TY: tropical year in days

LT: length and time conversion power of the sun (in mille joule per meter square)

NO: number of orbits per planet

NC: number of curvatures per orbit round the sun

NP: number of planets per cluster round the sun

SL: solar luminosity (in mille joule / meter square)

AT1: actual spinning time difference on a planet (in minutes)

TM1: time taken by a planet to cover its long orbital curvatures round the sun (in minutes)

TM2: time taken by a planet to cover its short orbital curvatures round the sun (in minutes)

AT2: actual rotational time difference on a planet (in seconds)

RTR: revolutional time relativity

TS: half of the time taken by a planet to cover its short axis (in hours)

TL: half of the time taken by a planet to cover its long axis (in hours)

OM: overall number of minutes per hour round the sun

NS: number of astronomical seasons round the sun.

AH: actual number of hours per day round the sun

RTL: Revolutionary time limit for all planetary bodies in our solar system(in days)

TCH: time conversion in hours per revolution
AM: Actual number of minutes per hour round the sun
TCD: time conversion in days per revolution
TP: time taken by a planet to make a complete rotation in minutes
OL: orbital length in thousand miles in which a planet swim round the sun (i.e LO = long orbit / SO = short orbit)
AL: Axial length in thousand miles on which a planet rotate (i.e LA = Long axis / SA = short axis)
STP: standard time taken by a planet to make a complete revolution (in days)
OL1: orbital length of a planet far away from the sun
OL2: Orbital length of a planet much closer to the sun
CTR: Celestial time relatively in hours
PWD: planetary weight dynamism (in pounds)
PWN: planetary weight per Newton (in kilogram)
TAR: tilt angle of reference (in degree)
EA: elliptical angle
OAR: overall angle of revolution in degree
ANC: astronomical number of cycles per orbit
NMC: number of months per year
NYM: number of years per millennium

**DERIVATION OF NEW ASTRONOMICAL CONSTANTS UNDERSTAND BY GMF**

Working From The First Dimension, We Have:
Input= [O$^I$ x  Q] A Slide per Letter
INPUT = GRAVITATION
DATA
O -26
I - 11 letters: 10, 9,8,7,6,5,4,3,2,1,0
Q-g = 6, r = 17, a =0, v = 21, i = 8, t =19, a = 0, t = 19, i =8, 0 = 14, n =13
COMPUTING
g[$26^{10}$ x 6] + r[$26^9$ x 17] + a[$26^8$ x 0] + v[$26^7$ x 21] + i[$26^6$    x 8] +t[$26^5$ x 19] + a[$26^4$ x 0] +t [$26^3$ x 19] +i[$26^2$ x 8] +
0[$26^1$ x 14] + n[$26^0$ x 13]
OUTPUT IN FIGURES
[939,475,501,888,625]-this is the Cardinal Value of the word 'gravitation' inside the F11
We are going to use the above cardinal value to derive our new astronomical constants understand by gravitational master formula.

The New Astronomical Constants
- GS = 2,055,853,070 Newton
- WS = 6,847439084 metric Tones
- DS = 45062 thousand Kilometers
- SL = 5,035,256 Mi/l2
- LT = 0.038md/m$^2$
- NC = 11
- NP = 6
- AM = 60 minutes
- DM = 80 minutes
- AS = 60 seconds
- AH = 24 hours
- NO = 1
- NYM = 1000 years
- NMY = 12months

## DEFINITION OF GRAVITATIONAL FORCE BASED ON NEW ASTRONOMICAL CONSTANTS [NAC]

Based on the new astronomical constants gravitational force is defined as the force of *repulsion* generated inside the *galactic center* which causes attraction and alignment, rotation and spinning, deflection and coordination between all the planetary bodies; and it was exemplary quantified to be *2, 055, 853, 070* Newton absorbed and distributed by the sun which has a weight of *6, 847, 439,084* metric tons inside the milky way.

**Galactic Centre** – is simply defined as the rotational center of the Milky Way

**Milky Way** – is simply defined as the spiral galaxy containing our solar system.

The following examples may give clear insight into the above definition of gravitational force

### I. Physics

Supposing, the milky way minimizes to a community transformer, the sun squeezes to a house meter, gravitational forces intensify to electric current, planetary bodies reduced to electrical appliances including *'Standing Fans and Light Bulbs'*. In what way would the house meter retrieve the quantity of gravitational forces used by the planetary bodies in the house?

### II. Astronomy

If the three types of gravitational force were to be diagrammatized based on their impact on the motion of planetary bodies, how does *the above definition of gravitational force* propose the *two basic types of motions* to be exhibited by a planetary body round the sun?

## III. Biology

Assuming, the above gravitational force is considered to be the only source of sunlight, then how will this force enhance the understanding of the process of germination involving the upward and downward movement of shoot and root, and the bending of growing stems and other plant parts toward source of light, a phenomenon which scientists defined generally as '*tropism*'

## THE ONLY ASTRONOMICAL VALUE UNDERSTANDS BY GMF

The astronomical record understands by the G.M.F is the tropical year which is simply defined as *the time taken by a planet to make a complete revolution round the sun. It was estimated using the "Armillary sphere" also known as the spherical astrolabe. This is the only record that can be used to retrieve all the basic astronomical features of a heavenly body inside the solar system*

### TROPICAL YEARS FOR NINE PLANETS INSIDE THE SOLAR SYSTEM

| PLANETS / MOONS / STARS | TROPICAL YEAR IN ROUND FIGURE |
|---|---|
| - Moon | 28 days |
| - Mercury | 88 days |
| - Venus | 225 days |
| - Earth | 365 days |
| - Mars | 687 days |
| - Jupiter | 4333 days |
| - Saturn | 10,756 days |
| - Uranus | 30,687 days |
| - Neptune | 60,190 days |
| - Pluto | 90,581 days |

#### WHAT DOES GMF CALCULATE?

Any time a tropical year was viewed through the GMF; the following astronomical features would be retrieved:

- Weight of a planet in metric tons (WP)
- Actual time taken by a planet to make a complete revolution in minutes[AT]
- Distance from the core to the surface of a planet in thousand kilometers (DP)

- Gravitational force acting on a planet in Newton (GP)
- Tropical year of a planet in days (TY)
- Actual Spinning time difference on a planet in minutes   (AT1)
- Actual Rotational time difference on a planet in seconds (AT2)
- Orbital length in thousand miles in which a planet revolve round the sun (OL)
- Axial length in thousand miles on which a planet rotate (AL)
- Standard time taken by a planet to make a complete revolution in days (STP)
- Celestial time relatively in hours (CTR)
- Planetary weight dynamism in pounds (PWD)
- Planetary weight per Newton in kilogram (PWN)
- Overall angle of revolution in degree(OAR)
- Tilt angle of reference in degree(TAR)
- Elliptical angle in degree[EA]
- Revolutionary time limit for all planetary bodies inside our solar system ,in thousand days (RTL)

## A VIEW OF SOLAR SYSTEM BASED ON 3 ASTRONOMICAL ERAS
### ➢ PREDICTION ERA

This is an era when the study of our solar system involved making positive and negative assertions about the future using the various positions of stars in the horizon. For example, the 12 zodiac signs, such as:

For example, the 12 zodiac signs, such as:

ARIES OR RAM:

There are thirteen stars that collectively form this shape. Its head is toward the west.

TAURUS:

In this figure there are thirty – three stars. Its tail is in the western direction and its head is to east, slightly inclined to the left. The horns point to the east.

GEMINI OR THE TWINS:

The stars in this shape are eighteen. This is a figure looking like two persons, formed from match sticks, holding each other arm. Their heads are to the north – east direction and their feet are to the south – west across to the midline of the sky.

CANCER OR CRAB: There are nine stars in all in this figure.

LEO OR LION:

Twenty – seven stars form this shape. Its forehead is to the west and it is the fifth constellation among the twelve.

## VIRGO OR VIRGIN:

This figure resembles a woman lying on her back on the midline of the sky. Its hand(s) point direct to the north and its legs to the east, slightly above the hands of Libra

## LIBRA:

There are eight stars forming this figure, neighboring the constellation Virgo.

## SCORPION:

Twenty – one stars form this figure

## SAGITTARIUS OR ARCHER:

There are thirty – one stars in this constellation. The figure looks like a man in a special output, , trying to draw an arrow in a bow.

## CAPRICORN:

Twenty – eight stars collectively form this shape

## AQUARIUS:

Forty – eight stars collectively form this shape; it is a figuring, apparently in a hurry, running while holding a container of water and in the process spilling some of it.

## PISCES OR FESHES:

There are thirty – four stars in this figure that has some features of fish.

These are the twelve constellations that ringed the night sky. They are sometimes referred to as the "Great Wheel of the Heaven", for they turn round ones in every twelve hours.

Note:

A constellation is a group of stars that collectively form the shape of a human being, animal, bird or the shape of any other recognizable object.

> **ESTIMATION ERA:**

This is an era when the study of our solar system is basically based on logical analyses to verify the assertions about the position of stars using the mathematical formulas e.g calculation of the radius of a star, the luminosity of a star, absolute magnitude of a star, apparent magnitude of star, the distance of a star and the temperature of a star.

## RADIUS OF A STAR

Is the distance from the center of a star to the surface or outside edge.

## LUMINOSITY OF A STAR

Is a measure of the total amount of energy radiated by a star per second. This is therefore the power output of a star.

## ABSOLUTE MAGNITUDE OF A STAR

Is the magnitude (brightness) of a star as it would be seen at a standard distance of 10parsecs (PC)

# APPARENT MAGNITUDE OF A STAR

Is the magnitude (brightness) it has as seen by an observer on earth.

# DISTANCE OF A STAR

It means hour for away a star is, from an observant on earth

# TEMPERATURE OF A STAR:

Is the degree of hotness or coldness based on the random speed of particles in a star.

The following mathematical formulas have gotten more popularity than the rest on the subject matter.

$L = R^2 \times T^4$

Where: L stands for luminosity of a star

R Stands for Radius of a Star

T Stand for surface temperature of a star

$$\text{Star's Flux} = \frac{Star's\ Luminosity}{[4x\ Star's\ distance]}$$

$$\text{Star's Distance} = \sqrt{\frac{Star's\ luminosity}{4\ x\ Star's\ flux}}$$

$m - M = 5\log \left[\frac{d}{10}\right]$

Where:

m- Stands for apparent Magnitude of a star

M – Stands for absolute magnitude of a star

D – Stands for distance to a star

v. $d = \frac{1}{P}$

Where:

d – Stands for distance in parsecs (PC)

p – Stands for parallax angle in arc seconds

## ➤ OBSERVATION ERA

This is an era when the study of our solar system is basically based on celestial visualization in order to standardize the astronomical estimations about a heavenly body using the observatory instruments, such as: Optical telescopes, Radio telescopes, the Gnomon and the Armillary sphere etc.

## OPTICAL TELESCOPES

Is a telescope that gathers light in the visible (or optical) part of the electromagnetic spectrum?

Note: Electromagnetic Spectrum: includes visible light, radio waves,

microwaves, infrared, ultraviolet, x – rays, and gamma rays.
There are three main types of optical telescopes, namely: refracting, reflecting, and catadioptric.

WHAT IS IT USED FOR?

To know:

How far away the star is

How old the star is

Whether or not the star is a part of a larger system and if there are any planets revolving around the star. Etc.

RADIOTELESCOPE:

is simply a telescope that is designed to receive radio waves (invisible light) from space. In its simplest form, it has three major components, namely: antenna, receiver and recorder.

WHAT IS IT USED FOR?

To enable objects invisible to the naked eye to be seen or photographed etc.

THE GNOMON:

is any object whose shadow is used to determine time.

WHAT IS IT USED FOR?

For finding the declination of the sun through the year

Estimation of angular direction of an object on the earth's surface

Telling time

THE ARMILLARY SPHARE (ASTROLABE)

is a miniature representation of celestial objects in the sky, depicted as a series of rings centered around a globe.

WHAT IS IT USED FOR?

To determine the time of sunrise and sunset

To determine the position of a star

To estimate the length of the tropical year

**DEFINITION OF SOLAR SPRING**

is the spiral arrangement of planetary bodies on the bases of weight and gravity, orbital and axial length.

TYPES OF SOLAR SPRING

There are basically two types of solar spring, namely:

Orbital and Axial spring

ORBITAL SPRING

is the spiral arrangement of planetary bodies on the bases of weight, gravity and orbital length.

AXIAL SPRING

is the spiral arrangement of planetary bodies on the bases of weight, gravity

and axial length.
## BASIC PARAMETERS OF SOLAR SPRING
WEIGHT OF A PLANET/GRAVITATIONAL FORCE
The weights of planets tend to decrease down the solar spring. Hence, the further a planet is from the sun, the less the chances of resisting higher gravitational force, which causes the planet to float beneath the solar system, and vice versa.
GRAVITATIONAL FORCE/PLANETARY DISTANCE
The gravitational force acting on planets tend to increase as the weight increases across the solar spring.
Hence, the closer a planets is to the sun, then the higher the gravitational force, also the faster the revolution and vice versa.
iii-ORBITAL AND AXIAL LENGTH/ROTATION
The orbital length of planets tends to increase as the distance between the planets increase down the solar spring, while the axial length tends to decrease as the distance between the planets increase down the solar spring. Hence, the smaller the orbital length, then the greater the axial length, also the slower the rotation and vice–versa

## CALCULATION OF THE KEY ASTRONOMIAL FEATURES OF OUR VARIOUS PLANETS INSIDE THE SOLAR SYSTEM
### ASTRONOMICAL QUESTION NO. 1
*Supposing our 12- zodiac sign would go round the sun 2 times in a day, use the Gravitational Master Formula to retrieve the key astronomical features of the 12- zodiac sign under the following guidelines:*
1. Weight of the zodiac signs in metric tons
2. Distance from the core to the surface of the zodiac signs in thousand kilometers.
3. Actual time taken by the zodiac signs to make a complete revolution in hours.
4. Gravitational force acting on the zodiac signs in Newton.
5. Circle in which the zodiac signs swim round the sun in thousand miles
6. Axial length on which the zodiac signs rotate in thousand miles.
7. Zodiac tilt angle of reference in degree
8. Zodiac elliptical angle.

Take the following 14 astronomical values to be constant
- GS = 2,055,853,070 Newton
- WS = 6,847439084 metric Tones
- DS = 45062 thousand Kilometers

AT = 24 /2
Therefore, AT = 12 hours per revolution

**4-Gravitational Force Acting On The Zodiac Signs Would Be Calculated As:**
From The General Unit Of The Formula, We Have:

$$GP = [\frac{GS}{AT} \ x \ \frac{WP}{DP} \ ] \frac{1}{4}$$

Data
GS = 2,055,853,070 Newton
AT = 1443.58 minutes
WP = 180195165368 metric tones
DP = 1185842.11 thousand kilometers

GP =(2055855070    x        180195765368) ¼
          1443.58                   1185842.11

= (1424135.18 x 151955.95) ¼
   (216405814205) ¼
= 5410145351.2
GP= 54101453551.2 Newton

**5 – Circular Curvature In Which The Zodiac Signs Revolve Round The Sun (In Thousand   Miles).**
From The Relation, We Have:
OL = (LO + SO)
Data
LO = 0.019
SO = 0.019
:. OL = (0.019+0.019)
OL = 0.038 thousand miles

6 – Axial Length On Which The Zodiac Rotate
 From The Relation:

(OL = TCH)
AL    TS +TL   for the axial length (overall length)

(LO = TCH)
LA    TS+TL    for long axis

(SO = TCH)
SA    TS+TL   for short axis

Note:
TS, [here] means half of the time taken by the zodiac signs to rotate on their short axis (in minutes)
TL,[ here] means half of the time taken by the zodiac signals to rotate on their long axis (in minutes).

From the first relation, we have

$$\frac{OL}{AL} = \frac{TCH}{TS+TL}$$

Where:

$$TS/TL =[ \frac{DP}{NSXOM}]1/12$$

Where:
For TS:

- DP = $\frac{[DP]}{SO}$ ½

- NS = $\frac{WP}{GP}$

-WP = $\frac{(WS)}{SO}$ ½

While for  TL:
- DP = $\frac{[DP]}{LO}$ ½

- NS = $\frac{[WP]}{GP}$ ½

- WP = $\frac{[WS]}{LO}$ ½

From TS, we have
- TS = $\frac{[DP]}{NSXOM}$ 1/12

- DP= $\frac{[DS]}{SO}$ ½

= $\frac{45062}{0.019}$ =1185842.11

DP=1185842.11thousands kilometers

WS = $\frac{WP}{GP}$

Where WP = $\frac{(WS)}{SO}$ ½

= $\frac{(6847439084)}{0.019}$ ½

WP = 180195765368

Therefore, NS = $\frac{180195765368}{5410145}$ = 3.3

∴ OM = 80 minutes

$$TS = [\frac{1185842.11}{3.33 \; x \; 80}] \; 1/12$$

$$= [\frac{11855842.11}{266.4}] \; 1/12$$

[4451.36] 1/12
[4451.36 x 0.083]
= 369.46 minutes
Therefore, our 'TS' in hours would be:
$\frac{369.46}{60}$ = 6.16 hours

As there is no difference between 'So' & 'LA' in length, hence the value of TL would not be different from the value of TS
Therefore, (TS + TL)
= (6.16 + 6.16)
= 12.32 => 12 hours[ in round figure]
Having obtained the value of [TS + TL], and the values of AT/TCH & OL were already at hand, then the value of AL would be simply retrieved from:
$$[\frac{OL}{AL} = \frac{TCH}{TS+TL}]$$

$$= \frac{0.038}{AL} = \frac{12}{12}$$
By cross multiplication, we have:
0.0138 x 12 = AL x 12
AL= $\frac{0.038 \; x \; 12}{12}$ therefore, AL = 0.038 thousand miles
From second relation, we have:
$$[\frac{LO}{LA} = \frac{TCH}{TS+TL}]$$

$$[\frac{0.019}{LA} = 12]$$
By cross Multiplication, we have:
0.019 x 12 = LA x 12
Therefore LA= $\frac{0.019 \; x \; 12}{12}$
LA = 0.019 thousand miles
= From third relation, we have:

$$\left[\frac{SO}{SA} = \frac{TCH}{TS+TL}\right] \text{Type equation here.}$$
$$= \left[\frac{0.019}{SA} = \frac{12}{12}\right]$$

AS Above:

Hence, SA = 0.019 thousand miles also.

7. Zodiac Tilt Angle of Reference

From the relation, we have:

$$TAR = [TCH \div \frac{LO}{SA}]$$

Where:

TCH = 12

LO = 0.019

SA = 0.019

$$TAR = [\ 12 \div \frac{0.619}{0.619}]$$

TAR = 12 ÷1

TAR = $12^0$

To retrieve the time required by our Zodiac signs to move at angle of $12^0$ , we should convert from degree to time using the following criteria.

$1^{st}$ – if the time taken per revolution by a heavenly body is less than 24 hours, then we should multiply the angle in degree by 2 to get our time in minutes.

$2^{nd}$ – if the time taken per revolution by a heavenly body is equal to or greater than 24 hours then we should square  the angle (in degree) to get our time in hours

For example:

Working from the first criterion, we have:

$12^0$ = [12 x 2]

$12^0$ = 24 minutes

Hence, our zodiac signs required 24 minutes to move at an angle of $12^0$ round the sun.

Similarly, if our zodiac signs move at an angle of $12^0$ in 24 minutes, then in 1 hour they would move at which angle round the sun.

-From the relation, we have:

$12^0$ = 24 minutes

X =  60 minutes[1 x 60minutes]

By cross multiplication, we have:

$12^0$ x 60 minutes = 24 minutes x  X

X = <u>$12^0$  x 60 minutes</u>
          24 minutes

$$\frac{12 \times 60}{24} = 12^0 \times 2.5 = 30^0$$

Generally, in 12 hours, is going to be

$12^0$ =24 minutes

X =  720[12 x 60 minutes]

Therefore, X  = $\underline{12^0 \times 720 \text{ minutes}}$

            24 minutes

$$\frac{12 \times 720}{24} = 12^0 \times 30$$

= $360^0$

Hence, the Overall Angle of Revolution required by our Zodiac signs is 360degrees

## 8 – Zodiac Elliptical Angle

From the relation, we have;

EA = OAR– [(ANC) x 360]

Where: ANC = $\frac{RND}{360}$

If OAR= 360, then ANC is going to be [(360/360) =1

EA = 360 – [1 x 360]

EA = 360 - 360

EA = 0

*Hence, there is no history of elliptical angle throughout the motion of zodiac signs round the sun.*

## ASTRONOMICAL QUESTION NO.2

*A heavenly body, the earth was discovered to be orbiting round the sun in every 365 days. Use the GMF to retrieve the key astronomical features of the planet under the following guidelines*:

1.Weight of the planet in metric tons (WP)

2.Distance from the core to the surface of the planet in thousand kilometers (DP)

3.Tropical year of the planet in days(TY)

4.Actual time taken by the planet to make a complete revolution in minutes(AT)

5.Gravitational force acting on the planet in Newton[GP].

6.Actual spinning time difference on the plane in minutes (AT1)

7.Actual rotational time difference on the planet in seconds(AT2)

8.Orbital length in thousand miles in which the planet revolve round the sun (OL)

9.Axial length in thousand miles on which the planets rotate (AL)

10. Standard time taken by the planet to make a complete     revolution in days (STP)

11.Planetary weight dynamism of the planet in pounds (PWD)

12.Planetary weight per Newton on the planet Earth in kilogram    (PWN)
13.Tilt angle of reference  of the planet in degree (TAR)
14.Elliptical angle of the planet in degree (EA)
15.Revolutional time limit (PTL) according to the planet in thousand days
Take the following 14 astronomical values to be constant

- GS = 2,055,853,070 Newton
- WS = 6,847439084 Metric Tones
- DS = 45062 thousand Kilometers
- SL = 5,035,256 Mi/l2
- LT = 0.038md/m$^2$
- NC = 11
- NP = 6
- AM = 60 minutes
- DM = 80 minutes
- AS = 60 seconds
- AH = 24 hours
- NO = 1
- NYM = 1000 years
- NMY = 12months

SOLUTION

-THE PLANET EARTH

**1.To Retrieve The Value Of WP, We Have:**

From the relation:

- WP= $\frac{WS}{LO}$

Data:

TY = 365 years

LT = 0.038mj/m$^2$

NO = 1

Therefore, LO = $[\frac{365 \times 0.038}{1}]$ ½

= $[\frac{13.87}{1}]$ ½

[13.87] ½

[13.87 x 0.5]

6.94Thousand miles

Therefore, WP = $\frac{WS}{LO}$

Data

WS = 6847439084

LO = 6.94

$WP = \frac{6847439084}{6.94}$

WP = 986662692.2 metric tons

## 2.To Retrieve The Value Of DP, We Have:

From the relation

$DP = \frac{DS}{LO}$

Data

DP = 45062 thousand kilometers

LO = 6.94 thousand miles

$DP = \frac{45062}{6.94}$

= 6493.08km

## 3.To Retrieve The Value Of TY, We Have:

From the relation:

$TY = [\frac{LO}{2LT} \times NO]$

Data

LO = 6.94

LT = 0.038

NO. = 1

Therefore, $TY = [\frac{6.94 \times 1}{\frac{1}{2}(0.038)}]$

$[\frac{6.94 \times 1}{0.19}]$

$= [\frac{6.94}{0.019}]$

= 365.26

Therefore, TY= 365days

## 4.To Retrieve The Value Of AT, We Have:

From the relation

$AT= [\frac{Wp}{DP} \times So] \frac{1}{2}$

Where:

$SO=[\frac{NP \times NC}{SL} \times \frac{WS}{DP}] \frac{1}{2}$ & $WP = \frac{WS}{LO}$

Data

NP = 6

NC = 11

SL = 5035256
WS = 6847439084
DP = 6493.08Km

$$SO = [\frac{6 \times 11}{5035256} \times \frac{684739084}{6493.08}] \ \frac{1}{2}$$

$$[\frac{66}{5035256} \times \frac{6847439084}{6493.08}] \ \frac{1}{2}$$

$[0.0000131 \times 1054574.88] \ \frac{1}{2}$

SO = $[13.81] \ \frac{1}{2}$
So = 6.91 thousand miles
= Therefore, by substitution, we have
From the relation-
AT = $[\frac{WP}{DP} \times SO] \ \frac{1}{2}$
Data
WP = 986662692.2
DP = 6493.08
SO = 6.91

$$AT = [\frac{9866626292.2}{6493.08} \times 6.91] \ \frac{1}{2}$$

AT = $[151956.04 \times 6.91] \ \frac{1}{2}$
= $[1050016.24] \ \frac{1}{2}$
$[1050016.24 \times 0.5]$
AT = 525008.12 minutes
TIME CONVERSION
-TCH = $\frac{AT}{Ah}$ =525008.12/60
=8750.14hours
-TCD =TCH/AM
$\frac{8750.14}{24}$ = 364.59 days make up a year on this plant
5.To Calculate GP, We Need To Work With The General Unit Of The Formula:
GP = $[\frac{GS}{AT} \times \frac{WP}{DP}] \ \frac{1}{2}$
Data
GS = 2055853070
AT = 525008.12
WP=986662692.2
DP=6493.08

$$GP = [\frac{2055853070}{525008.12} \times 151956.04] \ \frac{1}{2}$$

[3915.85 x 151956.04] ½
[595037059.2] ½
GP = 297,518,529.6 Newton.
## 6.To Retrieve The Value Of AT1 We Have:
From the relation
AT1 = [Tm1 + TM2]
Where:
= TM1 = [$\frac{WP}{DP}$ x LO] ¼

= TM2 = [$\frac{Wp}{DP}$ x SO] ¼
From TM1:
Data
$\frac{Wp}{DP}$ = 151956.04
LO= 6.94
[151956.04 x 6.94] ¼
= [1054574.92] ¼
[1054574.92 x 0.25]
= 263,643.73 minutes
TIME CONVERSION
TCH = $\frac{AT}{AM}$ = $\frac{263643.73}{60}$ = 4394.06

TCD = $\frac{TCH}{DH}$ = $\frac{4394.06}{24}$ = 183.09days make up half of a year on this planet
= From TM2, we have:
= [$\frac{WP}{DP}$ x SO] ½
Data
$\frac{WP}{DP}$ = 15195.04,
So = 6.91
TM2 = [151956.04 x 6.91] ¼
[1050016.24] ¼
= 262504.06 minutes
TIME CONVERSION
= TCH = $\frac{AT}{AM}$ = $\frac{262500.06}{60}$ = 4375.07 hours

-TCH = $\frac{TCH}{AH}$ = $\frac{4375.07}{24}$ = 182.29 days make up half of a year on this planet
Therefore AT1 = TM1 – TM2

= 263643.73 − 262504.06

= 1139.67 minutes

This implies that the spinning of the planet on its long orbital curvature is 1139.07 minutes longer than spinning on its short orbital curvatures round the sun. Hence, the planet moves faster on its long orbital curvature, than on its short orbital curvatures,; as the former is pushing the planet closer to the sun and vice-versa.

**7.To Retrieve The Value Of AT2, We Have:**

From the relation:

AT2 = [TS − TL]

Where:

$$TS = [\frac{DP}{NS \times OM}]\; \tfrac{1}{2}$$

$$TL = [\frac{DP}{NS \times OM}]\; \tfrac{1}{2}$$

From TS, we have

$$DP = \frac{DS}{SO}$$

$$NS = \frac{WP}{GP}, \text{ and } WP = \frac{WS}{SO}$$

= From DP. We have:

$$= \frac{45.062}{6.91} = 6521.27$$

From WP, we have;

$$\frac{6847439084}{6.91} = 990946321.9$$

If GP = 297518529.6, then the value of NS is going to be;

$$\frac{990946321.9}{297518529.6} = 3.33$$

Hence, the number of astronomical seasons round the sun is basically 3

Going back to the relation, we have;

$$TS = [\frac{DP}{NS \times OM}]\; \tfrac{1}{2}$$

Data

DP = 6521.27

NS = 3.33

OM = 80minutes

Therefore,

$$TS = [\frac{6521.27}{3.33 \times 80}]\; \tfrac{1}{2}$$

$$= [\frac{6521.27}{266.4}] \; ½$$

$= [24.48] \; ½ = 12.24$ hours

Hence, the time taken by the planet to make a complete rotation on its short – axis per day is 24.48 hours.

From TL, we have

$$TL = [\frac{DP}{NS \; x \; OM}] \; ½$$

Where

$$DP = \frac{DS}{LO} = 6493.08$$

$$NS = \frac{WP}{GP}$$

$$\frac{WP}{LO} = 98662692.2$$

If GP = 297518529.6, the NS is going to be

$$\frac{986662692.2}{297518529.6} = 3.32$$

Note: the value of GP need not to be broken on the basis of SO & LO, because the formula that work it out has taken care of that: 'AT always go along with SO', while 'WP was operated based on LO'

Going back to our relation, we have;

$$TL= [\frac{DP}{NS \; x \; OM}] \; ½$$

Data

DP = 6493.08

NS = 3.32

OM = 80minutes

$$[\frac{6493.08}{3.32 \; x \; 80}]½ = [24.45] \; ½ = 12.23 \text{ hours}$$

Hence, the time taken by the planet to make a complete rotation on its long axis per day is 24.45 hours

Therefore, AT2 = [TS – TL]

= [12.24 – 12.23]

= 0.01 hour

Which is equal to 3.6seconds[0.01 x 60 x 60]

Therefore,AT2 =3.6seconds

This implies that the rotation of the planet on its short axis is 3.6 seconds longer than rotation on its long axis. Hence the planet earth moves faster on its short axis, than on its long axis, as the former is pushing the planet closer to the sun and vice-versa.

**8.To Retrieve The Value Of 'OL', We Have:**
From the relation
OL = [LO + SO]
OL = 96.94 + 6.91
OL = 13.85 thousand miles
Hence, the planet swims around the sun in an orbit of length 13.85 thousand miles

**9.To Retrieve The Value Of Al, La And Sa We Have;**
From the relations
$[\frac{OL}{AL} = \frac{TCH}{TS+TL}]$ for AL [overall axial length]

$[\frac{LO}{LA} = \frac{TCH}{TS+TL}]$ for LA [long – axis]

$[\frac{SO}{SA} = \frac{TCH}{TS+TL}]$ for SA [short – axis]

From the first relation, we have:

$[\frac{OL}{AL} = \frac{TCH}{TS+TL}]$
Data
OL = 13.85 thousand miles
TCH = 8750.14 hours
(TS +TL) = 24 hours
AL = ?
Therefore AL = $\frac{OL \; x \; (TS+T2)}{TCH}$
AL = $\frac{13.85 \; x \; 24}{8750.14}$

= $\frac{332.4}{8750.14}$

AL = 0.038 thousand miles
From the second relation, we have
$[\frac{LO}{LA} = \frac{TCH}{TS+TL}]$
Data
LO = 6.94 thousand miles
TCH = 8750.14 hours
TS + TL = 24hours

LA = ?

$$LA = \frac{LO \times (TS+TL)}{TCH} = \frac{6.94 \times 24}{8750} = \frac{166.56}{8750.14} =$$

LA = 0.01904thousand miles

From the third relation, we have

$$[\frac{SO}{SA} = \frac{TCH}{TS+TL}]$$

Data

SO = 6.91 thousand miles

TCH = 8750.14 hours

(TS + TL) = 24 hours

SA = ?

Therefore, $SA = \frac{SO \times (TS+TL)}{TCH}$

$$= \frac{6.91 \times 24}{8750.14}$$

= SA = 0.01895 thousand miles

**10.To Retrieve The Value Of STP, We Have**

From the relation:

$$STP = [\frac{OL}{AL}] \times [\frac{TS+TL}{AT}]$$

Or

$$[\frac{LO}{LA} \times \frac{TS+TL}{AH}]$$

Or

$$[\frac{SO}{SA} \times \frac{TS+TL}{AH}]$$

From, $[\frac{OL}{AL} \times \frac{TS+TL}{AH}]$,we have

Data

OL = 13.85 thousand miles

AL =0.038 thousand miles

(TS + TL) = 24 hours

AH = 24 hours

Therefore, $STP= [\frac{13.85}{0.038} \times \frac{24}{24}]$

= 364.47 x 1

= 364.47 day

From, $[\frac{LO}{LA} \times \frac{TS+TL}{AH}]$, we have

Data

LO = 6.94

LA = 0.01904
TS + TL = 24
AT = 24
Therefore, STP = $[\frac{6.94}{0.01904} \times \frac{24}{24}]$
[364.50 x 1]
364.50 days
From $[\frac{SO}{SA} \times \frac{TS \, X \, TL}{AH}]$, we have
Data
LO = 6.91
SA = 0.01895
TS + TL = 24
AH = 24
Therefore STP = $[\frac{6.91}{0.01895} \times \frac{24}{24}]$
364.64 x 1
364.64days

**11.To Retrieve The Value Of PWD ,We Have:**
From the relation:
PWD = $[\frac{OL1}{OL2} \times ½]$
= taken the value of OL2 to be of the orbit in which the Earthly moon swim round the sun, which has a length 1.1 thousand miles.[ in one decimal place]
Data
OL1 (of the earth) – 13.85 thousand miles
OL2 (of the moon) – 1.1 thousand miles

PWD = $[\frac{13.85}{1.1} \times ½]$

[12.59 x ½]
[12.59 x 0.5] = 6 pounds
Hence, 6 pounds is the dynamism factor of earth to its moon.
12.To Retrieve The Value Of PWN, We Have:
From the relation:
-PWN = $[\frac{WP}{GP} \times AL]$
Data
WP = 986662692.2
GP = 297,518,529.6
AL = 0.038

PWN = $[\frac{986662692.2}{297,51829.6}$ x 0.038]

(3 x 0.038)

= 0.114kg

Or

From, PWN = $[\frac{WS}{GS}$ x AL]

Note: 'AL' is the axial length of a planet of interest.

Data

WP = 684739084

GS = 2055853070

AL = 0.038 [Of the earth]

Therefore, PWN = $[\frac{684749084}{2044853070}$ x 0.038]

(3 x 0,038)

= 0.114kg

Hence, one Newton is equivalent to 0.114kg on the planet earth.

**13. To Retrieve The Value Of TAR, We Have:**

From the relation

TAR= [TCH $\div \frac{LO}{SA}$]

Data

TCH = 8750.15 hours

LO = 6.94

SA = 0.01895

TAR = 8750.14 $\div \frac{6.94}{0.01895}$

TAR = 8750.14 $\div$ 366.23 = 23.89

Therefore, TAR = $23.89^0$

To retrieve the time required by the planet to move at angle of $23.89^0$, we should work from the second criterion, which says;

If the time taken per revolution by a heavenly body is equal to or greater than 24 hours ,then we square the angle [in degree]to get our time in hours

For example

$23.89^0 = (23.89)^2$

$23.89^0$ = 571hours

Hence, if 571hours is equivalent to $23.89^0$, then 24hours would be equivalent to how many degrees?

From the relation:

571hours = $23.89^0$

24hours = X (in degree)

$$X = \frac{23.89 \times 24}{571} = \frac{23.89 \times 24}{571}$$

$$\frac{573.36}{571} = 1^0$$

Therefore, 24hours is equivalent to $1^0$ on this planet

= Also, if 571 hours is equivalent to $23.89^0$, then 365 days would be equivalent to how many degrees

From the relation:

571 hours x X = $23.89^0$ x 365 x 24

$$X = \frac{23.89 \times 365 \times 24}{571} = \frac{209276.4}{571} =$$

X= $366.51^0$

Hence RND = 366.51degrees

Alternatively

If 24hours is equivalent to $1^0$, then 365days would be equivalent to how many degrees?

From the relation

= 24hour = $1^0$

365 x 24 = X

$$= \frac{365 \, X \, 24}{24}$$

X = $365^0$

## 14.To Retrieve The Value Of EA, We Have:

From the relation'

EA = OAR = [ANC] x 360

Where

$$ANC = \frac{RND}{360}$$

Data

OAR = 366.51

Therefore, ANC = $\frac{36651}{360}$ =1 cycle

EA = 366.51 – [1] x 360

EA = 366.51 – 360

EA = $6.51^0$ (e.g 366.51 – 6.51 = 360)

## *15.To Retrieve The Value Of RTL, We Have:*

From the relation;

RTL (in thousand days) = [ $\frac{OAR}{AL}$ x OL x NYM] provide that RTL - $\frac{OL}{TL}$ [in thousand days] = 0

From RTL, We have:

Data

OAR = 366.51 days
AL = 0.038 mj/m2
OL = 13.85 thousand miles
NYM = 1000 years
$$RTL = \frac{366.51}{0.038} \times 13.85 \times 1000$$

$$= \frac{5076163.5}{0.038} = 133583250 \text{ minutes}$$

$$= TCH = \frac{133.583250}{60} = 2226387.5$$

$$TCD = \frac{2226387.5}{24} = 92,000 \text{ thousand years i.e about 776days were reduced from}$$
92,766.15days)

Hence, for the planet to be able to determine the revolutional time limit of all the heavenly bodies inside our solar system, the value of its RTL must be equal to the value of $\frac{OL}{TL}$ (in thousand years

Let see how closer it may be;

DATA

OL=5035256

TL=0.038

$$= \frac{5035256}{0.238} = 132506736.8 \text{ minute}$$

$$TCH = \frac{132506736.8}{60} = 2208445.61 \text{ hours}$$

$$TCD = \frac{2208445.61}{24} = 92,000 \text{ thousand years i.e. about 19days were reduced}$$
from 92,018 .57days

Therefore, RTL =92,000 – 92,000 = 0

*Hence, the planet Earth has met the standard to determine the revolutional time limit for all the heavenly bodies in our solar system and hence, any planet with a tropical year greater than 92,000 thousand days should be considered to be among the exoplanets*

## ASTRONOMICAL QUESTION NO. 3

*Supposing an astronaut happen to land successfully on the moon at exactly 2.26 am and he was expected to send a signal transmission to his space base on the earth at exactly 9am, at what time the astronaut should get ready for the transmission of the signal?*

*Take the values of (TS + TL) of the planet closer to the sun to be 13.25 hours each.*

SOLUTION

From the relation, we have

$$CTR = \frac{TS+TL \text{ of a planet Much closer to the sun}}{TS+TL \text{ of a planet far away from the sun}}$$

Therefore;

$$\frac{Moon\ [TS + TL]}{Earth\ [TS + TL]}$$

For moon , we have;

TS = 13.25 hours

TL = 13.25 hours

[TS + TL) = [13.25 + 13.25] = 26.50 = 27 hours

For Earth [TS + TL], we have;

TS = 12.24 hours

TL = 12.23 hours

[TS + TL) = [13.25 + 13.25] = 26.50 = 27 hours

For Earth [TS + TL], we have;

TS = 12.24 hours

TL = 12.23 hours

[TS + TL] = [12.24 + 12.23] = 24.47 = 24hours

Therefore, $= \frac{27}{24}$ = 1.13 hours

Hence, 1 hour on the Earth is equivalent to 1.13 hour on the Moon. Therefore, time on the moon runs faster than the time on the Earth.

Then, if 1 hour on the earth is 1.13hour on the moon, 9am on the earth will be equal to;

From the relation:

1hour = 1.13hours

9hours = X

By cross multiplication, we have;

X x 1hour = 9hours x 1.13hours

$X = \frac{10.17}{1}$ = 10.17 hours on the moon

Hence, time on the earth will be;

$\frac{10.17}{1.13}$ = 9am

Therefore, the transmission should be sent at 10.17hours on the moon i.e in 7hours time when the corresponding time at 2.26am on the moon is 2am on the earth e.g $\frac{2.26}{1.13}$ = 2hours[and 9 – 2 = 7hours

## VERSATILE APPLICATION OF THE MACHINE: ONCOLOGY

### Critical Reasoning No.2

*You are asked to arrange the following basic risk factors of cancer in order of hazardousness, such as: sunlight, tobacco, beers, fake drugs, dusts, chemical carcinogens, fumes, and substances. Use any approach that doesn't contradict the rules of science.*

*Working From The First Dimension, We Have:*

INPUT=CANCER

Data

O- 26

I -5, 4, 3,2,1,0

Q-c=2,a=0,n=13,c=2,e=4,r=17

COMPUTING

$c[26^5 x2]+a[26^4 x0]+ n[26^3 x 13]+c[26^2 x2]+e[26^1 x 4]+r[26^0 x17]$

OUTPUT

[23992713]- This is `the cardinal quantitative value of cancer from which we can produce the cancer scoring system:

[23992713] -*[2-3-9-7-1]-Cancer Scoring Scale*

Interpretation of the scale

INPUT=CHEMICAL CARCINOGENS

COMPUTING

$c[26^7 x2]+h[26^6 x 7]+e[26^5 x 4]+m[26^4 x12]+i[26^3 x 8]+c[26^2 x2]+ a[26^1 x 0]+l[26^0 x 11] / + / c[26^{10} x 2 ]+a[26^9 x 0 ]+ r[26^8 x17]+c[26^7 x 2]+ i[26^6 x 8]+n[26^5 x13]+o[26^4 x 14]+g[26^3 x6]+ e[26^2 x 4]+n[ 26^1 x 13]+s[26^0 x 18]$

OUTPUT

[18279181971]+[285902947315172]

[285921226497143]

To verify the hazardousness of chemical carcinogens on cancer scoring system, the above value   must be reduced to one of the language of the machine

For example

[285921226497143]= [2+8+5+9+2+1+2+2+6+4+9+7+1+4+3]

[65]=[6+5]

[11]=[1+1]

[2]- [2-3-9-7-1]

It implies that chemical carcinogens are among the causes of cancer which the Most Hazardous.

INPUT=FUMES

COMPUTING

f[$26^4$ x  5]+u[$26^3$ x 20]+m[$26^2$x 12]+ e[$26^1$ x 4]+s[$26^0$ x 18]

OUTPUT

[2644634]

To verify the hazardousness of fumes on cancer scoring scale, the above value must be reduced to one of the language of the machine

For example

[2644634]+ [2+6+4+4+6+3+4]

[29]=[2 +9]

[11]

[2]- [2-3-9-7-1]

It implies that fumes are among the causes of cancer which are the Most Hazardous

INPUT=BEERS

COMPUTING

b[$26^4$ x  1]+e[$26^3$ x 4]+e[$26^2$x 4]+ r[$26^1$ x 17]+s[$26^0$ x 18]

OUTPUT

[ 530444]

To verify the hazardousness of beers on cancer scoring scale, the above value must be reduced to one of the language of the machine

For example

[ 530444]+ [5+3+0+4+4+4]

[20]=[2 +0]

[2]- [2-3-9-7-1]

It implies that beers are among the causes of cancer which are the Most Hazardous

INPUT=SMOKING

COMPUTING

s[$26^6$ x 18]+m[$26^5$ x 12]+o[$26^4$ x14]+k[$26^3$ x 10]+i[$26^2$x 8]+ n[$26^1$ x 13]+g[$26^0$ x 6]

OUTPUT

[5709639656]

To verify the hazardousness of smoking on cancer scoring scale, the above value   must be reduced to one of the language of the machine

For example

[5709639656] +[5+7+0+9+6+3+9+6+5+6]

[56]=[5+6]

[11]=[1+1]

[2]- [2-3-9-7-1]

This implies that smoking is among the causes of cancer which are the Most

Hazardous.
INPUT=TOBACCO
COMPUTING
$t[26^6 \times 18]+0[26^5 \times 12]+b[26^4 \times 14]+a[26^3 \times 10]+c[26^2 \times 8]+ c[26^1 \times 13]+o[26^0 \times 6]$
OUTPUT
[6036197402]
To verify the hazardousness of tobacco on cancer scoring scale, the above
value   must be reduced to one of the language of the machine
For example
[6036197402]= [6+0+3+6+1+9+7+4+0+2]
[38]
[11]
[2]- [2-3-9-7-1]
It implies that tobacco is among the causes of cancer which are the Most
Hazardous
INPUT=SUBSTANCES
COMPUTING
$s[26^9 \times 18 ]+ u[26^8 \times 20]+b[26^7 \times 1]+ s[26^6 \times 18]+t[26^5 \times 19]+a[26^4 \times 0]+n[26^3 \times 13]+ c[26^2 \times 2]+e[ 26^1 \times 4]+s[26^0 \times 18]$
OUTPUT
[1019214257833338]
To verify the hazardousness of substances on cancer scoring system, the
above value   must be reduced to one of the language of the machine
For example
[101921425783338   ]= [1+0+1+9+2+1+4+2+5+7+8+3+3+3+8]
[57]=[5+7]
[12]=[1+2]
[3]- [2-3-9-7-1]
It implies that substances are among the causes of cancer which are More
Hazardous.
INPUT=DUSTS
COMPUTING
$d[26^4 \times 3]+u[26^3 \times 20]+s[26^2 \times 18]+ t[26^1 \times 19]+s[26^0 \times 18]$
OUTPUT
[1735128 ]
To verify the hazardousness of dusts on cancer scoring scale, the above value
must be reduced to one of the language of the machine
For example
[ 1735128]+ [1+7+3+5+1+2+8]

[27]=[2 +7]
[9]- [2-3-9-7-1]
It implies that dusts are among the causes of cancer which are Much
Hazardous
INPUT=FAKE DRUGS
COMPUTING
f[$26^3$ x 5]+a[$26^2$x0] + k[$26^1$ x 10]+e[$26^0$ x 4] / + / d[$26^4$ x 3]+r[$26^3$ x17]+ u[$26^2$ x
20]+g[ $26^1$ x 6]+s[$26^0$ x 18]
OUTPUT
[88144]+[ 1683414]
[1771558]
To verify the hazardousness of fake drugs on cancer scoring system, the
above value   must be reduced to one of the language of the machine
For example
[1771558]= [1+7+7+1+5+5+8]
[34]=[3+4]
[7]- [2-3-9-7-1]
It implies that fake drugs are among the causes of cancer which are
Hazardous.
INPUT=SUNLIGHT
COMPUTING
s[$26^7$ x18]+u[$26^6$ x 20]+n[$26^5$ x 13]+l[$26^4$ x11]+i[$26^3$ x 8]+g[$26^2$ x 6]+ h[$26^1$ x
7]+t[$26^0$ x 19]
OUTPUT
[ 150910528177 ]
To verify the hazardousness of sunlight on cancer scoring system, the above
value   must be reduced to one of the language of the machine
For example
[ 150910528177 ]= [1+5+0+9+1+0+5+2+8+1+7+7]
[64]=[6+4]
[10]=[1+0]
[1]- [2-3-9-7-1]
It implies that sunlight is among the causes of cancer which Less Hazardous
***Critical Reasoning No.3***
*You are asked to arrange the following basic clinical manifestations of cancer*
*in order of severity, such as: thickness and redness of the skin, lump, nodule*
*and lesion .Your arrangement should enable early diagnosis of the disease.*
*Use any approach that doesn't contradict the rules of science.*
Working From The First Dimension, We Have

INPUT=CANCER
Data
O- 26
I -5, 4, 3,2,1,0
Q-c=2,a=0,n=13,c=2,e=4,r=17
COMPUTING
c$[26^5$ x2]+a$[26^4$ x0]+ n$[26^3$ x 13]+c$[26^2$ x2]+e$[26^1$ x 4]+r$[26^0$ x17]
OUTPUT
[23992713]- This is `the cardinal quantitative value of cancer from which we can produce the cancer scoring system:
-[2-3-9-7-1]-Cancer Scoring Scale
Interpretation of the scale
INPUT=NODULE
COMPUTING
n$[26^5$ x 13]+o$[26^4$ x 14]+d$[26^3$ x 3]+u$[26^2$x 20]+ I$[26^1$ x 11]+e$[26^0$ x 4]

OUTPUT
[160909753]
[29]
[11]
[2]- [2-3-9- 7-1]
It implies that of all the clinical manifestations of cancer nodule requires prompt consultation with a healthcare provider.
INPUT=LUMP
COMPUTING
I$[26^3$ x 11]+u$[26^2$x 20]+ m$[26^1$ x 12]+p$[26^0$ x 15]
OUTPUT
[207183]
[21]
[3]- [2-3-9- 7-1]
It implies that of all the clinical manifestations of cancer lump requires immediate consultation with healthcare provider.
INPUT=LESION
COMPUTING
I$[26^5$ x 11]+e$[26^4$ x 4]+s$[26^3$ x 18]+i$[26^2$x 8]+ o$[26^1$ x 14]+n$[26^0$ x 13]

OUTPUT
[132845193]
[36]

[9]- [2-3-9- 7-1]

It implies that of all the clinical manifestations of cancer, lesion requires an emergency consultation with healthcare provider.

INPUT=REDNESS

COMPUTING

$r[26^6 \times 17]+e[26^5 \times 4]+d[26^4 \times 3]+n[26^3 \times 13]+e[26^2 x4]+ s[26^1 \times 18]+s[26^0 \times 18]$

OUTPUT

[5300696302]

[34]

[7]- [2-3-9- 7 -1]

It implies that of all the clinical manifestations of cancer redness requires necessary consultation with healthcare provider.

INPUT=THICKNESS

COMPUTING

$t[26^8 \times 19]+ h[26^7 \times 7]+i[26^6 \times 8]+c[26^5 \times 2]+k[26^4 x10]+n[26^3 \times 13]+e[26^2 \times 4]+ s[26^1 \times 18]+s[26^0 \times 18]$

OUTPUT

[4026436788574]

[64]

[10]

[1]- [2-3-9- 7 - 1]

It implies that of all the clinical manifestations of cancer, thickness requires consultation with healthcare provider.

**Critical    Reasoning No. 4**

The following are considered to be the key diagnostic investigations for ruling out cancer, namely: genetic test, laboratory test, scanning, biopsy and marker test. You are to use this information to design a diagnostic catalog for establishing accurate diagnosis of cancer. Make sure that the catalog would not give any room for scientific dispute.

Working From The First Dimension, We Have:

INPUT=CANCER

Data

O- 26

I -5, 4, 3,2,1,0

Q-c=2,a=0,n=13,c=2,e=4,r=17

COMPUTING

$c[26^5 x2]+a[26^4 x0]+ n[26^3 \times 13]+c[26^2 x2]+e[26^1 \times 4]+r[26^0 x17]$

OUTPUT

[23992713]- This is `the cardinal quantitative value of cancer from which we

can produce the cancer scoring system:
[23992713] -[2-3-9-7-1]-Cancer Scoring Scale
Interpretation Of The Scale

**NEW CANCER DIAGNOSTIC CATALOG**
INPUT=SKIN BIOPSY
COMPUTING
s[$26^3$ x 18]+k[$26^2$ x 10]+ i[$26^1$ x 8]+n[$26^0$ x 13]+ b[$26^5$ x 1]+i[$26^4$ x 8]+o[$26^3$ x 14]+p[$26^2$ x 15]+ s[$26^1$ x 18]+y[$26^0$ x 24]
OUTPUT
[323349 ]  + [15793880 ]
[16117229 ]
[29]
[11]
[2]- [2-3-9- 7 - 1]
It implies that skin biopsy is the best diagnosis for ruling out cancer.
INPUT=SCANNING
COMPUTING
s[$26^7$ x 18]+c[$26^6$ x 2]+a[$26^5$ x 0]+n[$26^4$ x13]+n[$26^3$ x 13]+i[$26^2$ x 8]+ n[$26^1$ x 13]+g[$26^0$ x 6]
OUTPUT
[145196589648]
[66]
[12]
[3]- [2-3-9- 7 - 1]
It implies that after biopsy, scanning is the best diagnosis for ruling out cancer.
INPUT=LABORATORY TEST
COMPUTING
l[$26^9$ x 11]+a[$26^8$ x 0]+ b[$26^7$ x 1]+o[$26^6$ x 14]+r[$26^5$ x 17]+a[$26^4$ x 0]+t[$26^3$ x 19]+o[$26^2$x 14]+ r[$26^1$ x 17]+y[$26^0$ x 24] / +/t[$26^3$ x 19]+e[$26^2$x 4]+ s[$26^1$ x18]+t[$26^0$ x 19]
OUTPUT
[59737099427042] +[ 337135]
[59737099764177]
[81]
[9]- [2-3- 9 - 7 - 1]
It implies that after scanning, laboratory test is the best diagnosis for ruling out cancer.

INPUT=MARKER TEST
COMPUTING
m[$26^5$ x 12]+a[$26^4$ x 0]+r[$26^3$ x 17]+k[$26^2$x 10]+ e[$26^1$ x 4]+r[$26^0$ x 17] /
+/t[$26^3$ x 19]+e[$26^2$x 4]+ s[$26^1$ x18]+t[$26^0$ x 19]
OUTPUT
[142882185] +[ 337135]
[143219320]
[25]
[7]- [2-3- 9 - 7 - 1]
It implies that after laboratory test, marker test is the best diagnosis for
ruling out cancer.
INPUT=GENETIC TEST
COMPUTING
g[$26^6$ x 6]+e[$26^5$ x 4]+n[$26^4$ x 13]+e[$26^3$ x 4]+t[$26^2$x 19]+ i[$26^1$ x 8]+c[$26^0$
x 2] / +/t[$26^3$ x 19]+e[$26^2$x 4]+ s[$26^1$ x18]+t[$26^0$ x 19]
OUTPUT
[1907044206] +[ 337135]
[ 1907381341 ]
[37]
[10]
[1]- [2-3- 9 - 7 - 1]
It implies that after marker test, genetic test is the best diagnosis for
ruling out cancer.

## LINGUISTICS
### Critical Reasoning No. 5
The following fundamentals of _language_, such as: grammar, syntax, dictionary,
sentence, clause, tense, phrase, phonology, part of speech, figure of speech,
idiomatic expression ,punctuation marks, literature are be to arranged in a way
that the percentage that the knowledge of each will contribute towards
enhancing learning of English language is calculable. You either use
segmented code or integrated code, or any other method that doesn't
contradict the rules of science.
Working From The First Dimension, We Have:
INPUT=LANGUAGE
DATA:
O-26
I-7, 6,5,4,3,2,1,0
Q-l=11, a=0, n=13, g=6, u=20, a=0, g=6, e=4
COMPUTING

l[$26^7$ x11]+a[$26^6$ x 0]+n[$26^5$ x 13]+g[$26^4$ x6]+u[$26^3$ x 20]+a[$26^2$x0]+ g[$26^1$ x 6]+e[$26^0$ x 4]

OUTPUT

[88507463360]-this is the cardinal quantitative value of the word "language" when all the English words of 8 letters are arranged serially.

To derive our language scoring scale from the above quantitative value, we are to select the definite values only, each will appear only once and the arrangement is considered to be in order of priority whose hierarchy depends on the overall information observed from the available variables on scoring scale. It may be from left to right or vice versa

For example:

[88507463360]= [8-5-7-4-6-3]

Hence, our language scoring scale is going to be: [8-5-7-4-6-3] and will yield the following core ratios-6:5:4:3:2:1(e.g. :6 for 8, :5 for 5, :4 for 7, :3 for 4, :2 for 6 and :1 for 3)

Interpretation of the scale

INPUT=PHONOLOGY

COMPUTING

p[$26^8$ x 15]+ h[$26^7$ x 7]+o[$26^6$ x 0]+n[$26^5$ x 13]+o[$26^4$ x14]+l[$26^3$ x 11]+o[$26^2$x14]+ g[$26^1$ x 6]+y[$26^0$ x 24]

OUTPUT

[3193114519268]-this is the cardinal value of the word "phonology" when all the English words of 9 letters are arranged serially.

To verify the percentage that the knowledge of phonology will contribute towards enhancing learning of English language, the above value needs to be reduced to one of the language of the machine:

[3193114519268]= [3+1+9+3+1+1+4+5+1+9+2+6+8]

[53]=[5+3]

[8]- [8-5-7-4-6-3]

If the language scoring system will yield the following core ratios-6:5:4:3:2:1, then the percentage that the knowledge of phonology will contribute towards enhancing learning of English language would be calculated as:

$\frac{6}{21}$ X 100 =28.6%

INPUT=TENSE

COMPUTING

t[$26^4$ x 19]+e[$26^3$ x 4]+n[$26^2$x13]+ s[$26^1$ x 18]+e[$26^0$ x 4]

OUTPUT

[8762108]-this is the cardinal value of the word "tense" when all the English words of 5 letters are arranged serially.

To verify the percentage that the knowledge of tense will contribute towards enhancing learning of English language, the above value needs to be reduced to one of the language of the machine:

[8762108]= [8+7+6+2+1+0+8]

[23]=[2+3]

[5]- [8-5-7-4-6-3]

If the language scoring system will yield the following core ratios-6:5:4:3:2:1,then  the percentage that the knowledge of tense will contribute towards enhancing learning of English language would be calculated as:

$\frac{5}{21}$ X 100 =23.8%

INPUT=PART OF SPEECH

Let view each word separately:

-PART

COMPUTING

p[$26^3$ x 15]+a[$26^2$x 0]+ r[$26^1$ x 17]+t[$26^0$ x 19]

OUTPUT

[264101]-this is the cardinal value of the word "part" when all the English words of 4 letters are arranged serially.

-OF

COMPUTING

o[$26^1$ x 14]+f[$26^0$ x 5]

OUTPUT

[369]-this is the cardinal value of the word "of" when all the English words of 2 letters are arranged serially.

-SPEECH

COMPUTING

s[$26^5$ x 18]+p[$26^4$ x 15] + e[$26^3$ x 4]+e[$26^2$x 4]+ c[$26^1$ x 2]+h[$26^0$ x 7]

OUTPUT

[220792475]-this is the cardinal value of the word "speech" when all the English words of 6 letters are arranged serially.

To verify the percentage that the knowledge of part of speech will contribute towards enhancing learning of English language, the above values need to be reduced to one of the language of the machine:

[264101]+[369]+[220792475]

[221056954]= [2+2+1+0+5+6+9+5+4]

[34]=[3+4]

[7]- [8-5-7-4-6-3]

If the language scoring system will yield the following core ratios-6:5:4:3:2:1, then the percentage that the knowledge of part of speech will contribute towards enhancing learning of English language would be calculated as:

$\frac{4}{21}$ X 100 =19%

INPUT=DICTIONARY

COMPUTING

d[$26^9$ x 3]+i[$26^8$ x 8]+ c[$26^7$ x 2]+t[$26^6$ x 19]+i[$26^5$ x 8]+o[$26^4$ x14]+n[$26^3$ x 13]+a[$26^2$x 0]+ r[$26^1$ x 17]+y[$26^0$ x 24]

OUTPUT

[17981162251258]-this is the cardinal value of the word "dictionary" when all the English words of 10 letters are arranged serially.

To verify the percentage that the knowledge of dictionary will contribute towards enhancing learning of English language, the above value needs to be reduced to one of the language of the machine:

[17981162251258]

[58]

[13]

[4]- [8-5-7- 4 -6 - 3]

If the language scoring system will yield the following core ratios-6:5:4:3:2:1, then the percentage that the knowledge of dictionary will contribute towards enhancing learning of English language would be calculated as:

$\frac{3}{21}$ X 100 =14.3%


INPUT=GRAMMAR

COMPUTING

g[$26^6$ x 6]+r[$26^5$ x 17]+a[$26^4$ x 0]+m[$26^3$ x 12]+m[$26^2$x 12]+ a[$26^1$ x 0]+r[$26^0$ x 17]

OUTPUT

[205569089]-this is the cardinal value of the word "grammar" when all the English words of 7 letters are arranged serially.

To verify the percentage that the knowledge of grammar will contribute towards enhancing learning of English language, the above value needs to be reduced to one of the language of the machine:

[205569089]

[51]

[6]- [8-5-7- 4 - 6 - 3]

If the language scoring system will yield the following core ratios-6:5:4:3:2:1, then the percentage that the knowledge of grammar will contribute towards enhancing learning of English language would be calculated as:

$\frac{2}{21}$ X 100 =9.5%

INPUT=PUNCTUATION MARKS

COMPUTING

p[$26^{10}$ x 15]+ u[$26^9$ x 20]+n[$26^8$ x 13]+ c[$26^7$ x 2]+t[$26^6$ x 19]+u[$26^5$ x 20]+a[$26^4$ x 0]+t[$26^3$ 19]+i[$26^2$x 8]+ o[$26^1$ x 14]+n[$26^0$ x 13] / +/ m[$26^4$ x12]+a[$26^3$ x 0]+r[$26^2$x 17]+ k[$26^1$ x10]+s[$26^0$ x 18]

OUTPUT

[2228833431206993] +[5495482]

[2228833436702475]

To verify the percentage that the knowledge of punctuation marks will contribute towards enhancing learning of English language, the above value needs to be reduced to one of the language of the machine:

[2228833431206993]

[66]

[12]

[3]- [8-5-7- 4 -6 - 3]

If the language scoring system will yield the following core ratios-6:5:4:3:2:1, then the percentage that the knowledge of punctuation marks will contribute towards enhancing learning of English language would be calculated as:

$\frac{1}{21}$ X 100 =4.8%

TEST YOUR ABILITY:

Use the above examples and find out the percentage that the knowledge of the following fundamentals of language will contribute towards enhancing understanding of English language:

Syntax, literature, idiomatic expression, figure of speech, phrase, clause and sentence

Take the core ratios to be- 6:5:4:3:2:1

## MATHEMATICS

### Critical Reasoning No. 6

Your are provided with following collection of words of English ,you are to produce a skeletal frame of a computerized dictionary showing the arrangement of words not only alphabetically but also serially and generationally. You are to use any method of permutation that doesn't contradict the rules of science. The words to include consist of the following:

Oil, tree, bee, paid, ant, read, speak, pen, write, book, think, logic

Working with the segmented code, we have:

INPUT=OIL
COMPUTING
$o[26^2 \times 14]+ i[26^1 \times 8]+l[26^0 \times 11]$
OUTPUT
[9,683]
INPUT=TREE
COMPUTING
$t[26^3 \times 19]+r[26^2 \times 17]+ e[26^1 \times 4]+e[26^0 \times 4]$
OUTPUT
[345,544]
INPUT=BEE
COMPUTING
$b[26^2 \times 1]+ e[26^1 \times 4]+e[26^0 \times 4]$
OUTPUT
[784]
INPUT=ANT
COMPUTING
$a[26^2 \times 0]+ n[26^1 \times 13]+t[26^0 \times 19]$
OUTPUT
[357]

INPUT=PAID
COMPUTING
$p[26^3 \times 15]+a[26^2 \times 17]+ i[26^1 \times 8]+d[26^0 \times 3]$
OUTPUT
[263,851]
INPUT=BOOK
COMPUTING
$b[26^3 \times 1]+o[26^2 \times 14]+ o[26^1 \times 14]+k[26^0 \times 10]$
OUTPUT
[27,414]

INPUT=READ
COMPUTING
$r[26^3 \times 17]+e[26^2 \times 4]+ a[26^1 \times 0]+d[26^0 \times 3]$
OUTPUT
[301,499]

INPUT=WRITE
COMPUTING
$w[26^4 \times 22]+r[26^3 \times 17]+i[26^2 x 8]+ t[26^1 \times 19]+e[26^0 \times 4]$
OUTPUT
[10,358,170]
INPUT=SPEAK
COMPUTING
$s[26^4 \times 19]+p[26^3 \times 4]+e[26^2 x13]+ a[26^1 \times 0]+k[26^0 \times 10]$
OUTPUT
[8,491,922]
INPUT=THINK
COMPUTING
$t[26^4 \times 19]+h[26^3 \times 7]+i[26^2 x 8]+ n[26^1 \times 13]+k[26^0 \times 10]$
OUTPUT
[8,811,332]
INPUT=LOGIC
COMPUTING
$l[26^4 \times 11]+o[26^3 \times 14]+g[26^2 x6]+ i[26^1 \times 8]+c[26^0 \times 2]$
OUTPUT
[5,277,066]
INPUT=PEN
COMPUTING
$p[26^2 x 15]+ e[26^1 \times 4]+n[26^0 \times 13]$
OUTPUT
[10,257]

Hence, the skeletal frame for the computerized dictionary that will show arrangement of words both alphabetically, generationally and serially will be depicted as below:

| WORDS OF THREE LETTERS | | WORDS OF FOUR LETTERS | | WORDS OF FIVE LETTERS | |
|---|---|---|---|---|---|
| Serial No. | Words | Serial No. | Words | Serial No. | Words |
| 357 | Ant | 27,414 | Book | 5,277,066 | Logic |
| 784 | Bee | 263,851 | Paid | 8,491,992 | Speak |
| 9,684 | Oil | 301,499 | Read | 8,811,332 | Think |
| 10,257 | Pen | 345,544 | Tree | 10,358,170 | Write |

REFERENCES

1. Curtis Wilson (1989), "The Newtonian achievement in astronomy", ch. 13 (page 233 -274) in "Planetary from the Renaissance to the rise of astrophysics: 2A: Tycho Brahe to Newton", CUP 1989

2. Hook's 1674 statement in "An Attempt to prove the Motion of the Earth from Observation", is available in online facsimile here.

3. *Max Born* (1924), Einstein's Theory of Relativity (The 1962 Dover edition, page 348 lists a table documenting the observed and calculated values for the procession of the perihelion of Mercury, Venus, and Earth).

4. *Misner, Chales W.: Thorne,* S.; Wheeler John Archibad (1973). Gravitation, New York W.H. Freeman and Company. ISBN 0-7167 -0344-0 page 1049.

5. *"How Many Solar System Bodies".* NASA/JPL Solar System Dynamics. *Retrieved 20 April 2018.*

6. Philosophies Naturalis Principia Mathematica, First published on 5 July 1687

7. Page 309, in H W Turnbull [Ed], Correspondence Of Isaac Newton, Vol2[1676 – 1687], [Cambridge University Press, 1960]. Document □239.

8. Page 309, in H W Turnbull [Ed], Correspondence Of Isaac Newton, Vol2[1676 – 1687], [Cambridge University Press, 1960]. Document □235,24 November 1679.

9. Page 239 in Curtis Wilson [1989], the Newtonian Achievement Inastronomy'

10. Rahali (2016), The Discovery of English Vocabury Calculator Aijrstem, June – August, 2016 Issue Vol. 3